CAMBRIDGE TRACTS IN MATHEMATICS

General Editors

B. BOLLOBAS, H. HALBERSTAM, C. T. C. WALL

99   Some applications of modular forms

88   Some applications of modular
          forms

PETER SARNAK

Stanford University

Some applications of modular forms

The right of the
University of Cambridge
to print and sell
all manner of books
was granted by
Henry VIII in 1534.
The University has printed
and published continuously
since 1584.

CAMBRIDGE UNIVERSITY PRESS

Cambridge
New York   Port Chester
Melbourne   Sydney

CAMBRIDGE UNIVERSITY PRESS
Cambridge, New York, Melbourne, Madrid, Cape Town, Singapore, São Paulo

Cambridge University Press
The Edinburgh Building, Cambridge CB2 8RU, UK

Published in the United States of America by Cambridge University Press, New York

www.cambridge.org
Information on this title: www.cambridge.org/9780521402453

First published 1990
This digitally printed version 2008

A catalogue record for this publication is available from the British Library

ISBN 978-0-521-40245-3 hardback
ISBN 978-0-521-06770-6 paperback

To my parents Frieda and Leon

Preface

These notes are an expanded version of the Wittemore Lectures given at Yale in November 1988.[1] The material presented in the four chapters is more or less selfcontained. On the other hand, in the section at the end of each chapter called 'Notes and comments,' it is assumed that the reader is familiar with more advanced and sophisticated notions from the theory of automorphic forms. Some of the material presented here overlaps with a forthcoming book, 'Discrete groups, expanding graphs and invariant measures' by A. Lubotzky. The points of view, emphasis, and presentation in that book and the present notes are sufficiently different that we decided to keep the two works separate. The reader is encouraged to look at both treatments of the material.

Acknowledgements:

The author would like to thank A. Lubotzky, C. McMullen, N. Pippenger, J.P. Serre, and I. Vardi for their help in the preparation of this manuscript.

---

[1]The author would like to thank the Mathematics Department at Yale for their warm hospitality.

Contents

Introduction

Traditionally the theory of modular forms has been and still is, one of the most powerful tools in number theory. Recently it has also been successfully applied to resolve some long outstanding problems in seemingly unrelated fields. Our aim in this book is to describe three such applications, developing along the way the necessary methods and material from the theory of modular forms. Briefly, the problems we examine are the following:

(A) Ruziewicz's problem.

The problem is whether the Lebesgue measure λ on the n–sphere S^n is the unique rotationally invariant mean on $L^\infty(S^n)$. To put it in another context, an amenable topological group G is one which carries an invariant mean on $L^\infty(G)$. Uniqueness of such a mean is a difficult question and seldom discussed. Actually, Ruziewicz in the 1920's posed the problem of the uniqueness of rotationally invariant finitely additive measures defined on Lebesgue sets on S^n. The relation between these problems is that an invariant mean on $L^\infty(S^n)$ is a finitely additive measure ν which is moreover absolutely continuous with respect to Lebesgue measure λ, i.e., $\nu(E) = 0$ whenever $\lambda(E) = 0$. Tarski [Tar] has remarked that it follows from the Hausdorff–Banach–Tarski paradoxical decompositions of S^n, $n \geq 2$, that any rotationally invariant finitely additive measure on S^n, $n \geq 2$, must be absolutely continuous with respect to λ. Hence for $n \geq 2$ the invariant mean and Ruziewicz problems are equivalent. In Chapter 2 it is shown that for $n = 1$ the invariant mean λ is not unique while for $n \geq 2$ it is. The key in the solution for $n \geq 2$ is the construction of a remarkable finite set of rotations in $SO(n + 1)$. These rotations which are 'super ergodic' also have other interesting applications. The analysis leads to a classification of those compact connected Lie groups for which Haar measure is the unique mean.

(B) Ramanujan graphs.

In network theory, computer science, as well as in extremal graph theory, a basic problem is that of explicitly constructing highly connected sparse graphs. There are many measures of high connectivity. One which has turned out to be most important is the expansion property [A1] (see Chapter

1

3). Briefly, a bipartite graph on $|I| = n = |O|$ vertices (where I stands for inputs and O for outputs) and with kn edges (k fixed, $n \to \infty$) is an (n, k, c) expander if for every $A \subset I$ with $|A| \leq n/2$, we have that ∂A (= outputs joined to A) satisfies $|\partial A| \geq c|A|$. Here $c > 1$ is fixed and is called the expansion coefficient. The problem is to construct a family of (n, k, c) expanders, $n \to \infty$, with c as large as possible. It is easy to see from counting arguments (see Chapter 3) that expanders exist. However the explicit construction of such graphs is much more difficult and will be carried out in Chapter 3. We will give a simple explicit construction of such families which are almost as good as the 'random' graph in its expansion property. In some other respects these graphs are better than random and even optimal. We call these graphs that we construct Ramanujan graphs. We hope that after reading these notes the reader will agree with our choice of name. One other property of these Ramanujan graphs is that they are the first explicit examples of graphs of large girth (that is length of the shortest closed circuit) and large chromatic number (the least number of colors needed to color the vertices so that no two adjacent colors are the same). Here too, the existence of such graphs (with large girth and large chromatic number) was established by [Er] using counting arguments. This result was one of the early achievements of the theory of 'random graphs' [Bo2].

(C) Linnik Problem.

It is well known that $n > 0$ is a sum of three integer squares; $n = x^2 + y^2 + z^2$ iff $n \neq 4^a (8b - 1)$, (see Gauß [Ga]). Moreover, if n is large and is a sum of three squares then it can be represented as such in many ways. Linnik studied the question of the distribution of (x, y, z) as above, for large n. Developing elaborate and powerful methods [Li1] he proved under certain hypotheses that the projection of these solutions onto the unit sphere becomes equidistributed as $n \to \infty$. In Chapter 4 an unconditional proof of this fact is given. The key is a new estimate on Fourier coefficients of forms of half integral weight.

In fact what links problems A,B, and C is that they are all reduced to the problem of estimating the size of Fourier coefficients of modular forms. That is, they are reduced to the Ramanujan conjectures and their generalizations. Chapter 1 is devoted to developing the modular theory needed, as well as a powerful basic technique, via exponential sums, for estimating Fourier coefficients. We also develop some results on cancellations due to the sign of Kloosterman sums (see Theorem A.2.1 in Appendix 1.2 of Chapter 1) which gives progress towards the Linnik–Selberg conjecture, see 1.5.6.

Notes and historical comments

(A) For $n = 1$ the non-uniqueness of the invariant mean is due to Granierer [Gr] and Rudin [Ru]. The uniqueness for $n \geq 4$ is due to Margulis [Ma1] and Sullivan [Su]. The key to their solution being the use of Kazhdan's groups with 'property T' [Ka]. This method works only for $n \geq 4$. Drinfeld [Dr] using the adelic theory of automorphic forms, Jacquet–Langlands theory, as well as the solution to the Ramanujan conjectures due to Deligne [De], settled the remaining $n = 2$ and $n = 3$ cases, again showing the mean to be unique. Our proof of the uniqueness for $n \geq 2$ in Chapter 2 is new. For $n = 2$ it is based on the theory of Hecke operators on $L^2(S^2)$ developed in Lubotzky–Phillips–Sarnak [LPS1]. The solution presented also has the advantage of being effective and explicit in its construction of the crucial ε–good sets of rotations. In fact the construction is optimal and these rotations may be used to give optimally equidistributed sets of rotations, see the discussion at the end of Chapter 2.

(B) The first explicit construction of an expanding family of graphs is due to Margulis [Ma2], though his construction does not yield an expansion coefficient. Many related constructions followed and the work of Alon–Millman [AM] and Alon [A1] crystallized the relation of expanders to eigenvalues of the adjacency matrix. The graphs presented in Chapter 3 are due to Lubotzky–Phillips–Sarnak [LPS2,LPS3]. They give the best explicit expanders known and are also optimal in related aspects. Margulis [Ma3,Ma4] has independently discovered similar constructions. For applications of expander graphs to nonblocking networks and to constructions of super–concentrators see Pippenger [Pi1].

(C) The unconditional solution to the Linnik problem and especially the estimation of Fourier coefficients of forms of 1/2–integral weight is due to Iwaniec [Iw2]. We will follow his method closely. The corresponding result for indefinite forms, the distribution of reduced binary quadratic forms and the corresponding estimation of Fourier coefficients of half integral weight Maaß forms was recently done by Duke [Du].

The basic method of estimating Fourier coefficients exploited in these notes is due to Kloosterman [Kl] and Petersson [Pe]. Selberg [Se] has given an insightful discussion of this and other methods. Theorem A.2.1 which gives the cancellation in signs of Kloosterman sums is due to Kuznietsov [Ku]. The simple proof given here is due to Goldfeld and Sarnak [GS].

Chapter 1

Modular Forms

1.1 Introduction

Ramanujan in his paper [R1] made two deep conjectures about the coefficients $\tau(n)$ of the function Δ

$$\Delta(q) = q \prod_{n=1}^{\infty} (1-q^n)^{24} = \sum_{n=1}^{\infty} \tau(n) q^n . \tag{1.1.1}$$

The first was the multiplicativity of the coefficients

$$\tau(mn) = \tau(m)\,\tau(n) \quad \text{if } (n,m) = 1 ,$$

the second an estimate

$$|\tau(n)| \le d(n)\, n^{11/2} , \tag{1.1.2}$$

where $d(n)$ is the number of divisors of n,

$$d(n) = \sum_{d|n} 1 . \tag{1.1.3}$$

The first was proved by Mordell [Mo] and marked the beginning of Hecke's theory of Hecke operators [H]. The second was proved by Deligne [De] and is one of the crowning achievements of mathematics. It is interesting to note that Ramanujan was interested in the bound (1.1.2) and related ones because $\tau(n)$ and related quantities appear as remainder terms in certain asymptotics. For example, if $r_{2s}(n)$ is the number of representations of n as a sum of $2s$ squares then he notes that

$$r_{2s}(n) = \delta_{2s}(n) + \varepsilon_{2s}(n) , \tag{1.1.4}$$

where $\delta_{2s}(n)$ is an arithmetical function involving sums over divisors of n and $\varepsilon_{2s}(n)$ is the remainder. For $s \ge 2$ the order of magnitude of $\delta_{2s}(n)$ is

n^{s-1}, while the analog of Conjecture (1.1.2) (for $s = 12$ the $\varepsilon(n)$ is essentially $\tau(n)$) is

$$\varepsilon_{2s}(n) = O_\epsilon\left(n^{\frac{s-1}{2}+\epsilon}\right) \qquad \text{for } \epsilon > 0. \qquad (1.1.5)$$

That is, $\delta_{2s}(n)$ is a strikingly good approximation to $r_{2s}(n)$. For example, if $s = 2$ then, as Jacobi showed, the remainder term is zero. In fact

$$r_4(n) = \delta_4(n) = 8 \sum_{\substack{d|n \\ 4 \nmid d}} d. \qquad (1.1.6)$$

Jacobi's proof of (1.1.6) used theta functions. We now develop the theory of theta functions and modular forms, which will provide the natural setting and explanation of the above considerations of Ramanujan.

1.2   Modular forms of integral weight

Let $\mathbf{H} = \{z| \operatorname{Im}(z) > 0\}$ denote the upper half plane. The group

$$SL(2,\mathbf{R}) = \left\{ \begin{pmatrix} a & b \\ c & d \end{pmatrix} \;\middle|\; a,b,c,d \in \mathbf{R}, \; \begin{vmatrix} a & b \\ c & d \end{vmatrix} = 1 \right\}$$

acts on $\mathbf{H}$ by linear fractional transformations $z \mapsto \frac{(az+b)}{(cz+d)}$. Let

$$\Gamma(1) = \left\{ \begin{pmatrix} a & b \\ c & d \end{pmatrix} \in SL(2,\mathbf{R}) \;\middle|\; a,b,c,d \in \mathbf{Z} \right\}$$

be the usual modular group. $\Gamma(1)$ is a discrete subgroup of $SL(2,\mathbf{R})$ and acts discontinuously on $\mathbf{H}$. A fundamental domain for this action is the familiar region $\mathcal{F}$ (see Figure 1.1). Thus $\Gamma(1)\backslash\mathbf{H}$ has one cusp, viz. ∞, with corresponding stabilizer of ∞ being the subgroup Γ_∞ of $\Gamma(1)$

$$\Gamma_\infty = \left\{ \begin{pmatrix} 1 & n \\ 0 & 1 \end{pmatrix} \in SL(2,\mathbf{R}) \;\middle|\; n \in \mathbf{Z} \right\}. \qquad (1.2.1)$$

The Riemann surface $\Gamma(1)\backslash\mathbf{H}$ has genus zero as is easily checked. We begin with the definition of holomorphic modular forms for $\Gamma(1)$ of even integral weight k.

Definition 1.2.1. A *holomorphic modular form of weight k* (an even integer) for $\Gamma(1)$ is a holomorphic function on $\mathbf{H}$ satisfying

(*i*) $f(\gamma z) = (cz + d)^k f(z) \qquad \gamma = \begin{pmatrix} * & * \\ c & d \end{pmatrix} \in \Gamma(1)$

(*ii*) $f(z)$ is bounded in the cusp of $\Gamma(1)\backslash\mathbf{H}$.

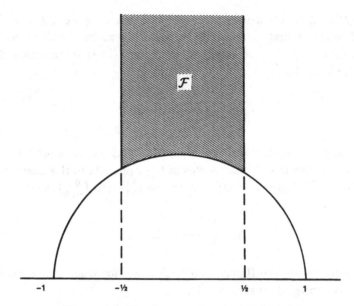

Figure 1.1: Fundamental region $\mathcal{F}$ for $\Gamma(1)$

The second condition asserts that f is holomorphic 'at ∞'. The local variable at ∞ is $q = e^{2\pi i z}$, and since $f(z+1) = f(z)$, f has a Fourier development

$$f(q) = \sum_{n=-\infty}^{\infty} a_n q^n . \qquad (1.2.2)$$

Condition (ii) ensures that $a_n = 0$ for $n < 0$, that is f is holomorphic in $|q| < 1$. The numbers a_n in $(1.2.2)$ are called the Fourier coefficients of f (at the cusps at ∞). If $w = \gamma z$ then

$$\frac{dw}{dz} = (cz + d)^{-2} \qquad \text{or } dw = (cz + d)^{-2} \, dz .$$

Hence a modular form of weight k for $\Gamma(1)$ corresponds to a meromorphic differential of weight $k/2$ on $\Gamma(1) \backslash \mathbf{H}$, that is $f(z) \, (dz)^{k/2}$ is such a differential (it may have a pole at ∞ in the variable q). With this remark we can easily compute the dimension of the finite dimensional space of modular forms of weight k for $\Gamma(1)$ by appealing to the Riemann–Roch theorem. We will not need (for the most part) the exact dimension, it will suffice to note that in this situation and the more general ones that follow, that the space of forms is finite dimensional. This follows by considering $\int_{\partial \mathcal{F}} (f'/f)(z) dz$ which, from the transformation rules of (i) is independent of f.

Definition 1.2.2. A modular form for $\Gamma(1)$ is called a *cusp form* if the coefficient a_0 in $(1.2.2)$ is zero.

Clearly the space of cusp forms of weight k is a subspace of the modular forms of weight k and corresponds to those f's which vanish at ∞. An example of a cusp form of weight $k = 12$ for $\Gamma(1)$ is the function $\Delta(q)$, $q = e^{2\pi i z}$, of (1.1.1), i.e.,

$$\Delta(z) = \sum_{n=1}^{\infty} \tau(n)\, e(nz)$$

(we use the notation $e(z) = e^{2\pi i z}$ for $z \in \mathbf{C}$). That $\Delta(z)$ is a modular form is by no means obvious. We prove this in the Appendix to this chapter. We note that clearly $\Delta(z+1) = \Delta(z)$ and since $\left(\begin{smallmatrix}1&1\\0&1\end{smallmatrix}\right)$ and $\left(\begin{smallmatrix}0&1\\-1&0\end{smallmatrix}\right)$ generate $\Gamma(1)$ it is sufficient to show that

$$\Delta(-1/z) = z^{12}\, \Delta(z). \tag{1.2.3}$$

We can now state the Ramanujan conjectures for holomorphic modular forms of even integral weight on $\Gamma(1)$.

Ramanujan Conjectures 1.2.3. *Let $f(z)$ be a cusp form of weight k for $\Gamma(1)$ then*

$$a_n = O_\epsilon(n^{(k-1)/2+\epsilon}) \qquad \text{for all } \epsilon > 0,$$

where the a_n's are the Fourier coefficients of f.

For the function $\Delta(z)$ this agrees with (1.1.2) (at least as far as the exponent) since $d(n) = O_\epsilon(n^\epsilon)$.

Our assumption that f be a cusp form for $\Gamma(1)$ is too restrictive. Forms invariant by certain subgroups of $\Gamma(1)$ will play a central role in the later chapters. Let $\Gamma(N)$ denote the principal congruence subgroup (of $\Gamma(1)$) of level N, where $N \geq 1$ is an integer. It is defined by

$$\Gamma(N) = \left\{ \begin{pmatrix} a & b \\ c & d \end{pmatrix} \in \Gamma(1) \,\middle|\, \begin{pmatrix} a & b \\ c & d \end{pmatrix} \equiv \begin{pmatrix} 1 & 0 \\ 0 & 1 \end{pmatrix} \pmod{N} \right\} \tag{1.2.4}$$

Any subgroup $\Gamma(N) \subseteq \Gamma \subseteq \Gamma(1)$ is called a congruence subgroup. Of special interest is $\Gamma_0(N)$ defined by

$$\Gamma_0(N) = \left\{ \begin{pmatrix} a & b \\ c & d \end{pmatrix} \in \Gamma(1) \,\middle|\, N|c \right\}. \tag{1.2.5}$$

If Γ is a congruence subgroup of $\Gamma(1)$ of index m then clearly a fundamental domain for Γ may be taken as

$$\mathcal{F}_\Gamma = \bigcup_{j=1}^{m} \gamma_j\, \mathcal{F}, \qquad m = [\Gamma(1), \Gamma]$$

for suitable $\gamma_j \in \Gamma(1)$. The set of rationals in $\mathbf{R}$ will break up under the action of Γ on $\mathbf{R}$ into say r equivalence classes of cusps $p_1, \ldots, p_r$. That is, for $i \neq j$, $\gamma p_i \neq p_j$ for $\gamma \in \Gamma$ and any $p \in \mathbf{R} \cup \{\infty\}$ is of the form $p = \gamma p_j$ for some $\gamma \in \Gamma$. We will always take $p_1 = \infty$. Since $\Gamma(1)$ has only one cusp, there is, for each p_j, a $\gamma_j \in \Gamma(1)$ such that $\gamma_j p_j = \infty$. To define a modular form of weight k for Γ we require, as before, that f be holomorphic on $\mathbf{H}$ and

$$f(\gamma z) = (cz + d)^k f(z), \quad \gamma \in \Gamma.$$

Also we need to ensure that $f(z)$ is holomorphic at each cusp. To examine f in the cusp p_j, it is convenient to introduce a local variable at p_j by mapping $p_j \mapsto \infty$ with the γ_j above. If $w = \gamma_j z$ then in the variable w, f is transformed into the form

$$F(w) = f(\gamma_j^{-1} w) \left(\frac{dz}{dw} \right)^{k/2}. \tag{1.2.6}$$

$F(w)$ is a form of weight k for $\Gamma_j = \gamma_j \Gamma \gamma_j^{-1} \subset \Gamma(1)$. We say that $f(z)$ is holomorphic at p_j if $F(w)$ is holomorphic at ∞, the Fourier coefficients of f at p_j are those of F at ∞. With respect to this, notice that $(\Gamma_j)_\infty \subset \left\{ \begin{pmatrix} 1 & n \\ 0 & 1 \end{pmatrix} \mid n \in \mathbf{Z} \right\}$ is of finite index, so that the Fourier development of F is of the form

$$F(z) = \sum_{n=0}^{\infty} a_n e\,(n\,z/M) \tag{1.2.7}$$

for some integer $M \geq 1$.

Definition 1.2.4. *A modular function of weight k for Γ a congruence subgroup of $\Gamma(1)$, is a holomorphic function $f(z)$ on $\mathbf{H}$ satisfying*

(i) $\quad f(\gamma z) = (cz + d)^k f(z)$

(ii) $\quad f(z)$ *is holomorphic at each cusp p_j*

(iii) $\quad f$ *is a cusp form if its zeroth Fourier coefficient is zero in each cusp.*

We denote the space of modular forms (cusp forms) of weight k for Γ by $\mathcal{M}_k(\Gamma)$, $(\mathcal{S}_k(\Gamma))$. Notice that a modular form for Γ is automatically one for $\Gamma' \subset \Gamma$ and similarly for cusp forms. We can now state the more general Ramanujan conjectures.

Ramanujan's Conjectures 1.2.5. *Let $f(z)$ be a holomorphic cusp forms of weight k for Γ (a congruence subgroup) then*

$$a_n = O_\epsilon(n^{(k-1)/2+\epsilon}) \qquad \text{for all } \epsilon > 0.$$

Here the a_n are the Fourier coefficients of f at any cusp.

In this situation we need only check the Ramanujan conjecture for the
cusp at ∞, since if $f(z)$ is a cusp form for $\Gamma(N)$ then $f(\gamma_j z)$ is also one for
$\Gamma(N)$, where $\gamma_j \in \Gamma(1)$.

To end this paragraph we make some comments about forms of odd weight
k. Though we will not make any use of these forms here they are of great
interest in number theory [DS] and in Physics [Sar]. The condition $f(\gamma z) =$
$(cz + d)^k f(z)$, when k is odd would be impossible if $\gamma = \left(\begin{smallmatrix} -1 & 0 \\ 0 & -1 \end{smallmatrix}\right)$. Hence,
for odd k, either we must assume that $\left(\begin{smallmatrix} -1 & 0 \\ 0 & -1 \end{smallmatrix}\right) \notin \Gamma$ or that we have an odd
character χ of Γ into $\{z|\ |z| = 1\}$, i.e., $\chi(-I) = -1$. Considering only
congruence groups, we also assume that $\ker \chi$ is a congruence subgroup. A
modular form of odd weight k is then as before a holomorphic function is
H, satisfying
$$f(\gamma z) = \chi(\gamma)\,(cz + d)^k f(z)\,, \qquad \gamma \in \Gamma\,.$$

Such a form with, say $k = 1$, corresponds to a tensor $f(z)\,(dz)^{1/2}$ on the
surface $\Gamma \backslash H$, that is a holomorphic spinor [Sar].

1.3   Theta functions and modular forms of 1/2-integral weight

One of the most important methods and certainly the one most relevant to
our applications, for constructing modular forms is via theta functions. We
begin with the classical theta function

$$\tilde{\theta}(z) = \sum_{m=-\infty}^{\infty} e^{i\pi m^2 z}\,. \tag{1.3.1}$$

This series is absolutely convergent for $\mathrm{Im}\,(z) > 0$ and it clearly satisfies

(i) $$\tilde{\theta}(z + 2) = \tilde{\theta}(z)\,.$$

A second transformation rule for $\tilde{\theta}(z)$ under $z \mapsto -1/z$ is derived from
the Poisson summation formula and is the basic ingredient in verifying the
transformation properties of any type of theta function.

Poisson summation 1.3.1. *Let f be a Schwartz class function on $\mathbf{R}^n$ then*

$$\sum_{m \in \mathbf{Z}^n} f(m) = \sum_{m \in \mathbf{Z}^n} \hat{f}(m)\,,$$

where

$$\hat{f}(\xi) = \int_{\mathbf{R}^n} f(x)\,e(-\langle x, \xi \rangle)\,dx\,.$$

In particular, since the Fourier transform of $e^{-\pi x^2}$ is $e^{-\pi \xi^2}$ we have

$$\sum_{n=-\infty}^{\infty} e^{-\pi n^2 y} = \frac{1}{\sqrt{y}} \sum_{n=-\infty}^{\infty} e^{-\pi n^2/y}, \qquad y > 0.$$

Hence

(ii) $$\tilde{\theta}(-1/z) = \sqrt{-iz}\,\tilde{\theta}(z).$$

(Here and elsewhere $\sqrt{}$ will be the usual branch, positive on $\mathbf{R}^+$.) This gives us the second transformation rule for $\tilde{\theta}(z)$ and hence implicitly also the transformation of $\tilde{\theta}$ under the group generated by $\begin{pmatrix} 1 & 2 \\ 0 & 1 \end{pmatrix}$ and $\begin{pmatrix} 0 & 1 \\ -1 & 0 \end{pmatrix}$. This is not explicit enough for our purposes . Let $\begin{pmatrix} a & b \\ c & d \end{pmatrix} \in SL(2,\mathbf{Z})$, $a \equiv 0 \pmod 2$, $d \equiv 0 \pmod 2$; we examine the behavior of $\tilde{\theta}$ under $\begin{pmatrix} a & b \\ c & d \end{pmatrix}$. Let $c > 0$ ($c < 0$ is dealt with similarly);

$$\tilde{\theta}\left(\frac{az+b}{cz+d}\right) = \tilde{\theta}\left(\frac{a}{c} - \frac{1}{c(cz+d)}\right)$$

$$= \sum_{m \,(\mathrm{mod}\, c)} e^{i\pi m^2 a/c} \sum_{t=-\infty}^{\infty} e^{-i\pi\left(\frac{m}{c}+t\right)^2 \frac{c}{cz+d}}$$

$$= (ic)^{-1/2}(cz+d)^{1/2} \sum_{m\,(\mathrm{mod}\, c)} e^{i\pi m^2 a/c} \sum_{\nu=-\infty}^{\infty} e^{2\pi i\, \frac{m\nu}{c}+i\nu^2\pi\left(z+\frac{d}{c}\right)}, \qquad (1.3.2)$$

where we have applied Poisson summation to the inner sum. Now

$$\sum = \sum_{m\,(\mathrm{mod}\, c)} e^{i\pi m^2 a/c + 2\pi i m\nu/c}$$

$$= \sum_{m\,(\mathrm{mod}\, c)} e\left(\frac{\alpha m^2 + m\nu}{c}\right) \qquad \text{where } \alpha 2 = a.$$

Since $(2\alpha)\, d \equiv 1 \pmod c$ and also c is odd, we can change variable

$$m = r\bar{\alpha},$$

where here and elsewhere $\alpha\bar{\alpha} \equiv 1 \pmod c$. The last then becomes

$$\sum = \sum_{r\,(\mathrm{mod}\, c)} e\left(\frac{\bar{\alpha}(r^2 + r\nu)}{c}\right) = \sum_{r\,(\mathrm{mod}\, c)} e\left(\frac{\bar{\alpha}(r + \bar{2}\nu)^2 - \bar{\alpha}\bar{4}\nu^2}{c}\right)$$

i.e.,

$$\sum = e\left(\frac{-\nu^2 \bar{4}\bar{\alpha}}{c}\right) \beta(\alpha,c)$$

where

$$\beta(\alpha, c) = \sum_{r \,(\mathrm{mod}\, c)} e\left(\frac{\overline{a}\, r^2}{c}\right). \tag{1.3.3}$$

Substituting for $\sum$ in (1.3.2) gives

$$\tilde{\theta}\left(\frac{az + b}{cz + d}\right) = (-ic)^{-1/2}\,(cz + d)^{1/2}\,\beta(\alpha, c)\,\tilde{\theta}(z). \tag{1.3.4}$$

Now β is a standard Gauß sum and may be explicitly evaluated [Da1]

$$\sum_{n=0}^{N-1} e\left(\frac{n^2}{N}\right) = \begin{cases} N^{1/2} & \text{if } N \equiv 1 \,(\mathrm{mod}\, 4) \\ i\, N^{1/2} & \text{if } N \equiv 3 \,(\mathrm{mod}\, 4) \end{cases}$$

Thus putting $-1/z$ for z in (1.3.4) and using this evaluation of the Gauß sum yields

$$\tilde{\theta}(\gamma\, z) = \left(\frac{2\, c}{d}\right) \varepsilon_d^{-1}\,(cz + d)^{1/2}\,\tilde{\theta}(z) \tag{1.3.5}$$

for any $\gamma \in SL(2, \mathbf{Z})$ with $c \equiv 0 \,(\mathrm{mod}\, 2)$, $b \equiv 0 \,(\mathrm{mod}\, 2)$, where $\varepsilon_d = 1$ or i depending on whether $d \equiv 1$ or $3 \,(\mathrm{mod}\, 4)$, and the symbol $(\div)$ is the Legendre symbol extended as follows, see Shimura [Sh]; for b odd

(i) $\left(\frac{a}{b}\right) = 0$ if $(a, b) \neq 1$,

(ii)   if b is an odd prime then $\left(\frac{a}{b}\right)$ is the usual Legendre symbol,

(iii) $b > 0$ then $\left(\frac{a}{b}\right)$ is a character $(\mathrm{mod}\, b)$,

(iv) $a \neq 0$, $\left(\frac{a}{b}\right)$ is a character $(\mathrm{mod}\, 4a)$

$$\left(\frac{a}{-1}\right) = 1 \text{ if } a > 0 \text{ and is } -1 \text{ if } a < 0, \quad \left(\frac{0}{\pm 1}\right) = 1.$$

If we set $\theta = \tilde{\theta}(2\, z) = \sum_{-\infty}^{\infty} e(m^2 z)$ then we conclude from (1.3.5):

Proposition 1.3.2.

$$\theta(\gamma\, z) = j(\gamma, z)\,\theta(z) \quad \text{for } \gamma \in \Gamma_0(4)$$

where

$$j(\gamma, z) = \left(\frac{c}{d}\right) \varepsilon_d^{-1}(cz + d)^{1/2}.$$

θ is our fundamental modular form of $1/2$ integral weight, in fact its multiplier $j(\gamma, z)$ is used to define these forms.

Definition 1.3.3. *Let* $4|N$. *A modular form* $f(z)$ *of weight* k *(k a $1/2$ integer) for* $\Gamma_0(N)$ *is a holomorphic function* $f(z)$ *on* $\mathbf{H}$ *satisfying*

(i) $$f(\gamma z) = (j(\gamma, z))^{2k} f(z) \quad \text{for } \gamma \in \Gamma_0(N),$$

(ii) $$f(z) \quad \text{is holomorphic at each cusp.}$$

This definition is consistent with the previous ones when k is even. Again let $\mathcal{M}_k(\Gamma)$, $S_k(\Gamma)$, denote the spaces of modular and cusp forms of weight k (they are clearly finite dimensional). It is also convenient to allow more generally forms transforming by

$$f(\gamma z) = \chi(\gamma)\, (j(\gamma, z))^{2k} f(z), \qquad \gamma \in \Gamma_0(N),$$

where $\chi\begin{pmatrix} a & b \\ c & d \end{pmatrix} = \chi(d)$ is a multiplicative character (mod N).

The above construction of $\theta(z)$ is a special case of the following construction due to Schoenberg [Sc] and Pfetzer [Pf]. We follow Shimura's treatment [Sh].

Let A be an $n \times n$ positive definite integral matrix. Let N be an integer such that $N A^{-1}$ is also integral. $P(x)$ denotes a spherical harmonic relative to A, that is, a homogeneous polynomial of degree $\nu \geq 0$ for which

$$\sum_{i,j} \tilde{a}_{ij} \frac{\partial^2 f}{\partial x_i \partial x_j} = 0, \qquad (1.3.6)$$

where $[\tilde{a}_{ij}] = A^{-1}$. For $h \in \mathbf{Z}^n$ let

$$\tilde{\theta}(z, h, N) = \sum_{m \equiv h \,(\mathrm{mod}\, N)} P(m)\, e\left(\frac{(^t m\, A\, m)\, z}{2\, N^2}\right). \qquad (1.3.7)$$

Clearly this series converges and defines a holomorphic function on $\mathbf{H}$. Using the method described for $\tilde{\theta}(z)$ one can show (see Shimura [Sh]) that:

(i) $$\tilde{\theta}(-1/z, h, N) = (-i)^\nu\, D^{-1/2}(-iz)^{k/2} \sum_{\substack{k \,(\mathrm{mod}\, N) \\ A\, k \equiv 0 \,(\mathrm{mod}\, N)}} e\left(\frac{^t k\, A k}{N^2}\right) \theta(z, k, N)$$

where $D = \det A$ and $k = n + 2\nu$.

(ii) $$\tilde{\theta}(z + 2, h, N) = e\left(\frac{^t h\, A h}{N^2}\right) \tilde{\theta}(z, h, N).$$

The second statement is obvious, the first follows from the Poisson summation.

(iii)
$$\tilde{\theta}(\gamma z, h, N) = e\left(\frac{ab\, ^t h\, A\, h}{2 N^2}\right)\left(\frac{\det A}{d}\right)\left(\frac{2c}{d}\right)^n \varepsilon_d^{-n}\, (cz + d)^{k/2}\, \tilde{\theta}(z, ah, N)$$

for $\gamma = \begin{pmatrix} a\,b \\ c\,d \end{pmatrix} \in SL(2, \mathbf{Z})$ and $b \equiv 0 \pmod 2$, $c \equiv 0 \pmod{2N}$.

The $\theta(z, h, N)$ above clearly give us an abundance of modular forms. We consider some examples.

Example 1: With $n = 1$, $N = 1$, $P(m) = m^\nu$, $\nu = 0, 1$, then

$$\tilde{\theta}(z) = \sum_{m=-\infty}^{\infty} m^\nu \, e(|m|^2 z/2),$$

which gives our old θ–function. More generally, if $\nu = 1$ and ψ is the Dirichlet character $\pmod 4$ with $\psi(-1) = -1$, then from the above

$$\theta(z, \psi) = \sum_{m=-\infty}^{\infty} \psi(m) \, m \, e(m^2 z) \tag{1.3.8}$$

is a cusp form for $\Gamma_0(8)$ of weight $3/2$.

Example 2: If

$$A = \begin{pmatrix} 1 & & \cdots & 0 \\ & 1 & & \\ \vdots & & \ddots & \\ 0 & & & 1 \end{pmatrix}$$

is the $n \times n$ identity matrix, $N = 1$ and P is a spherical harmonic of degree ν,

$$\tilde{\theta}_P(z, 0, 1) = \sum_{m \in \mathbf{Z}^n} P(m) \, e(|m|^2 z/2)$$

satisfies the transformation rule (iii) above. Hence setting

$$\theta_P(z) = \text{const.} \, \tilde{\theta}_P(2z) = \sum_{m \in \mathbf{Z}^n} P(m) e(|m|^2 z)$$

$$= \sum_{t=0}^{\infty} \left(\sum_{|m|^2 = t} P(m) \right) e(tz) \tag{1.3.9}$$

we find that

$$\theta_P \left(\frac{az + b}{cz + d} \right) = (j(\gamma, z))^k \, \theta_P(z). \tag{1.3.10}$$

Moreover (i) above shows that θ_P is holomorphic at every cusp and if $\nu > 0$ it is clear from the definition of $\tilde{\theta}$ and (i) above that θ_P is a cusp form. That is, we have

$$\theta_P \in S_k(\Gamma_0(4)) \text{ for } \nu \geq 1, \; k = \frac{n}{2} + \nu$$

$$\tag{1.3.11}$$

while for $P(x) \equiv 1, \quad \theta_1 \in \mathcal{M}_{n/2}(\Gamma_0(4))$.

Note that

$$\theta_1(z) = \sum_{\nu=0}^{\infty} r_n(\nu)e(\nu z)$$

using the notation of (1.1.4).

Example 3: Let A be an integral 4×4 positive matrix as above. Then $Q(x) = {}^t x \, A \, x$ is a quaternary form. Let

$$r_Q(\nu) = \#\{m \in \mathbf{Z}^4 | \, Q(m) = \nu\} \,.$$

If $N \, A^{-1}$ is integral then the above considerations especially (iii) above show that

$$\theta(z) = \sum_{m \in \mathbf{Z}^4} e(({}^t m \, A \, m) \, z) = \sum_{\nu=0}^{\infty} r_Q(\nu) \, e(\nu z) \qquad (1.3.12)$$

is in $\mathcal{M}_2(\Gamma(N))$.

We now formulate the Ramanujan conjectures for $1/2$–integral weight. Some care must be exercised in the formulation since for example, the cusp form $\theta(z, \psi)$ in (1.3.8) for $\Gamma_0(8)$ of weight $3/2$ has

$$a_\nu = \sqrt{\nu} \qquad \text{for } \nu = m^2 \,.$$

This is clearly not $O\left(\nu^{\frac{k-1}{2}+\epsilon}\right)$ for $\epsilon < 1/4$. Thus $\theta(z, \psi)$ fails to satisfy the Ramanujan bound. Hence we must either keep away from such θ–functions of one variable or, which can be shown to be the same thing (see the notes to Chapter 4), assume that n is square free (or has a fixed square factor).

Ramanujan Conjecture for half integral weight 1.3.4. *Let $f(z)$ be a holomorphic cusp form of weight k (where k is half an odd integer $k \geq 3/2$) for $\Gamma_0(N)$; if*

$$f(z) = \sum_{n=1}^{\infty} a_n e(nz)$$

then for square free n, $a_n = O_\epsilon(n^{(k-1)/2+\epsilon})$ for $\epsilon > 0$.

We will return to this conjecture later and in Chapter 4 make some progress towards its proof. Suffice it to say here that unlike the even weight case, Conjecture 1.3.4 is far from being completely solved at present.

We end this section by pointing out what will be termed the 'trivial' bound for the Fourier coefficients of a cusp form.

Proposition 1.3.5. *Let $f \in S_k(\Gamma)$ with Fourier coefficients a_n then*

$$a_n = O(n^{k/2}) \,.$$

Proof: Since $y(z) = \text{Im}\,(z)$ satisfies $y(\gamma\,z) = y(z)/|cz + d|^2$ we see that

$$F(z) = |f(z)|\,y^{k/2}$$

is Γ invariant. Also, since f vanishes at each cusp, it follows that $F(z)$ is bounded on $\mathbf{H}$, say $|F(z)| \leq M$. Now

$$a_n = e^{-2\pi n y} \int_0^1 e(-nx)\, f(x + iy) dx\,.$$

Hence $|a_n| \leq M\,e^{2\pi n y}\,y^{-k/2}$. Choosing $y = 1/n$ yields the claim. $\qquad\square$

The Ramanujan conjectures give the strongest possible bound for a_n since it can be shown (see Selberg [Se]) that

$$\sum_{n\leq X} \frac{|a_n|^2}{n^{k-1}} \sim c\,X\,,$$

with $c > 0$.

Any bound on the Fourier coefficients beyond the trivial one is significant since it typically leads to a solution of the problem at hand, even if not the optimal one. A notable exception to the last is the Linnik problem as is explained in Chapter 4. In Section 1.5 we will present a non-trivial bound of $O_\epsilon(n^{k/2-1/4+\epsilon})$ for $\epsilon > 0$ for the coefficients.

1.4   Eisenstein series

In addition to theta series, Eisenstein series also produce modular forms though strictly noncuspidal ones. Moreover this construction and the forms so produced are well understood.

Assume first that $k > 2$. For each p_j of $\Gamma\backslash\mathbf{H}$ we define an Eisenstein series $E_k^{(j)}(z)$. We do this first for $p_1 = \infty$; the other Eisenstein series are constructed using their local variable in the same way. We assume further (without loss of generality) that

$$\Gamma_\infty = \Gamma \bigcap \left\{ \begin{pmatrix} 1 & t \\ 0 & 1 \end{pmatrix} \middle|\, t \in \mathbf{R} \right\} = \left\{ \begin{pmatrix} 1 & n \\ 0 & 1 \end{pmatrix} \middle|\, n \in \mathbf{Z} \right\}\,. \qquad (1.4.1)$$

Define

$$E_k^{(\infty)}(z) = \sum_{\gamma\in\Gamma_\infty\backslash\Gamma} (j(\gamma, z))^{-2k}\,. \qquad (1.4.2)$$

The first issue is that of convergence. This is best understood through the 'spectral' Eisenstein series, for $\text{Re}\,(s) > 1$

$$E(z, s) = \sum_{\gamma\in\Gamma_\infty\backslash\Gamma} \frac{y^s}{|cz + d|^{2s}} = \sum_{\gamma\in\Gamma_\infty\backslash\Gamma} (\text{Im}\,(\gamma\,z))^s\,. \qquad (1.4.3)$$

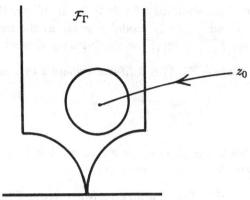

Figure 1.2: Non–Euclidian ball of radius δ

Let $z_0 \in \mathcal{F}_\Gamma$ with $B(z_0, \delta) \subset \mathcal{F}_\Gamma$ where $B(z_0, \delta)$ is a non–Euclidian ball radius δ about z_0 (see Figure 1.2). For $s > 1$ the series in (1.4.3) converges absolutely and

$$E(z_0, s) \leq \frac{c(s)}{\mathrm{vol}\,(B(z_0, \delta))} + O(y_0^s). \tag{1.4.4}$$

To see this note that for z_0' with $y_0' \leq 1$ (say) then for some C, $y_0' \leq Cy$, for $z \in B(z_0, \delta)$, and δ small. Hence for real s

$$\sum_{\gamma \in \Gamma_\infty \backslash \Gamma} y(\gamma\, z_0)^s \leq C^s \sum_{\gamma \in \Gamma_\infty \backslash \Gamma} \frac{1}{\mathrm{vol}\,(B(\gamma\, z_0, \delta))} \int_{B(\gamma\, z_0, \delta)} y^s \frac{dx dy}{y^2},$$

where $dx dy / y^2$ is the area element of the hyperbolic plane. The sets $B(\gamma\, z_0, \delta)$, $\gamma \in \Gamma_\infty \backslash \Gamma$ are disjoint and all lie in the intersection $\{z|\, -1/2 < \mathrm{Re}\,(z) \leq 1/2\} \cap \{z|\, \mathrm{Im}\,(z) \leq T\}$. Hence for $s > 1$

$$E(z_0, s) \ll \frac{C^s}{\mathrm{vol}\,(B)} \int_0^T \int_0^1 y^s \frac{dx dy}{y^2} \ll \frac{C^s}{\mathrm{vol}\,(B(z_0, \delta))}.$$

The above holds for $y_0 \leq 1$. In general we get at most an extra term coming from $\gamma = $ identity giving: For $\mathrm{Re}\,(s) > 1$

$$E(z_0, s) \ll \frac{C(\sigma)}{\mathrm{vol}\,(B(z_0, \delta))} + y^\sigma, \tag{1.4.5}$$

where $\sigma = \mathrm{Re}\,(s)$ and $B(z_0, \delta) \subset \mathcal{F}_\Gamma$. Returning to $E_k^{(\infty)}$ we have

$$y^{k/2} |E_k^\infty(z)| \leq \sum_{\gamma \in \Gamma_\infty \backslash \Gamma} |cz + d|^{-k} y^{k/2} = E(z, k/2). \tag{1.4.6}$$

Hence $E_k^{(\infty)}(z)$ converges absolutely for $k > 2$. It follows that $E_k^{(\infty)}(z)$ is holomorphic in $\mathbf{H}$ and is clearly modular as far as the transformation property goes. In fact $E_k^{(\infty)} \in \mathcal{M}_k(\Gamma)$ as the following shows:

Proposition 4.1. *Assume* $\left(\begin{smallmatrix} -1 & 0 \\ 0 & -1 \end{smallmatrix}\right) \in \Gamma$ *(if not, replace 2 by 1 below)*;

$$\lim_{\substack{z \to \infty \\ z \in \mathcal{F}_\Gamma}} E_k^{(\infty)}(z) = 2$$

and $E_k^{(\infty)}(z)$ *vanishes at the other cusps* $p_2, \ldots, p_r$. *Similarly* $E_k^{(p_j)}(z)$ *has value 2 at* p_j *and vanishes at the other cusps.*

Proof: We prove it for $E_k^{(\infty)}(z)$. The absolutely convergent Eisenstein series will (except for the terms involving the cosets $\Gamma_\infty I$, $\Gamma_\infty(-I)$) tend to zero as $z \to \infty$. It follows that $E_k^{(\infty)}(\infty) = 2$. Next we examine the behavior of $E_k^{(\infty)}(z)$ as $z \to p_j$, another cusp. Suppose for definiteness that $p_j = 0$. From (1.4.6) and (1.4.5) we have

$$|E_k^\infty(z)| \leq y^{-1-k/2} \qquad \text{as } z \to 0.$$

In terms of the local variables, we have

$$|F(z)| = |z^{-k} E_k(-1/z)| \leq |z|^{-k} y^{k/2+1} \to 0 \qquad \text{as } z \to \infty \text{ since } k > 2.$$

This proves the Proposition. $\qquad\qquad\qquad\qquad\qquad\qquad\qquad\qquad\qquad\square$

Corollary 1.4.2. *The Eisenstein series* $E_k^{(p_j)}(z)$, $j = 1, \ldots, r$ *span an* $r-$ *dimensional subspace of* $\mathcal{M}_k(\Gamma)$. *Moreover, each* $g \in \mathcal{M}_k(\Gamma)$ *has a unique representation*

$$g = e + h$$

where e *is in the space spanned by the Eisenstein series and* $h \in S_k(\Gamma)$ *(i.e., a cusp form).*

Proof: Follows directly from the Proposition. $\qquad\qquad\qquad\qquad\qquad\square$

Fourier development of (1.4.3). The Fourier coefficients of the Eisenstein series defined above are easily computed. What we find is that the nth coefficient is of the form: a sum over the divisors d of n, of powers of d – 'a divisor sum'. In particular, the coefficients are elementary arithmetical functions of n.

Consider for example the case of $\Gamma_0(N)$ and $E_k^{(\infty)}(z)$. If

$$E_k(z) = \sum_{\gamma \in \Gamma_\infty \backslash \Gamma_0(N)} (j(\gamma, z))^{-2k} = \sum_{m=0}^{\infty} a_m \, e(mz)$$

then

$$a_n = \int_0^1 E_k(z)e(-nz)dz \qquad \text{(more precisely } \int_{i\alpha}^{i\alpha+1} \dots dz \text{ with } \alpha > 0)$$

$$= 2\int_0^1 e(-nz)dz + 2 \sum_{\substack{c>0 \\ \left(\begin{smallmatrix} a\,b \\ c\,d \end{smallmatrix}\right)\in\Gamma_\infty\backslash\Gamma/\Gamma_\infty}}$$

$$\sum_{m\in\mathbf{Z}} \int_0^1 \left(j\left(\gamma\begin{pmatrix} 1\,m \\ 0\,1 \end{pmatrix}, z\right)\right)^{-2k} e(-nz)dz$$

$$= 2\delta_{0,n} + 2 \sum_{\substack{c>0 \\ \left(\begin{smallmatrix} a\,b \\ c\,d \end{smallmatrix}\right)\in\Gamma_\infty\backslash\Gamma/\Gamma_\infty}}$$

$$\sum_{m\in\mathbf{Z}} \int_0^1 \left(\left(\tfrac{c}{d}\right) \varepsilon_d^{-1}(c(z+m)+d)^{1/2}\right)^{-2k} e(-nz)dz$$

$$= 2\delta_{0,n} + 2 \sum_{\substack{c>0 \\ \left(\begin{smallmatrix} a\,b \\ c\,d \end{smallmatrix}\right)\in\Gamma_\infty\backslash\Gamma/\Gamma_\infty}} \left(\left(\tfrac{c}{d}\right) \varepsilon_d^{-1}\right)^{-2k} \int_{-\infty}^{\infty} (cz+d)^{-k} e(-nz)dz$$

$$= 2\delta_{0,n} + 2 \sum_{\substack{c\equiv 0 \ (\mathrm{mod}\ N) \\ c>0}} c^{-k}$$

$$\sum_{\substack{d\ (\mathrm{mod}\ c) \\ (d,c)=1}} \left(\left(\tfrac{c}{d}\right) \varepsilon_d^{-1}\right)^{2k} \int_{-\infty}^{\infty} \left(z+\tfrac{d}{c}\right)^{-k} e(-nz)dz$$

$$= 2\delta_{0,n} + 2 \sum_{\substack{c\equiv 0 \ (\mathrm{mod}\ N) \\ c>0}} c^{-k}$$

$$\sum_{\substack{d\ (\mathrm{mod}\ c) \\ (d,c)=1}} \left(\tfrac{c}{d}\right)^{-2k} \varepsilon_d^{2k}\, e\left(\tfrac{nd}{c}\right) \int_{-\infty}^{\infty} z^{-k} e(-nz)dz$$

Therefore

$$a_n = 2\delta_{0,n} + 2n^{k-1} \left(\int_{-\infty}^{\infty} z^{-k} e(-z)dz \right) \sum_{\substack{c\equiv 0 \ (\mathrm{mod}\ N) \\ c>0}} c^{-k}$$

$$\sum_{\substack{d\ (\mathrm{mod}\ c) \\ (d,c)=1}} \left(\tfrac{c}{d}\right)^{-2k} \varepsilon_d^{2k}\, e\left(\tfrac{nd}{c}\right). \tag{1.4.7}$$

For k half an odd integer the exponential sum in (1.4.7) is a little more complicated and will be dealt with in Chapter 4 together with related sums. For k even the sum in (1.4.7) is

$$\sum_{\substack{c \equiv 0 \ (\text{mod } N) \\ c > 0}} c^{-k} \sum_{\substack{d \ (\text{mod } c) \\ (d,c)=1}} \left(\frac{c}{d}\right)^{-2k} \varepsilon_d^{2k} \, e\left(\frac{nd}{c}\right). \tag{1.4.8}$$

The inner sum is easily evaluated as was done by Ramanujan [R2] and gives

$$\sum_{\substack{c \equiv 0 \ (\text{mod } N) \\ c > 0}} c^{-k} \sum_{\substack{d|c \\ d|n}} \mu\left(\frac{c}{d}\right) d, \tag{1.4.9}$$

where μ is the Möbius function. For example if $N = 1$ this gives

$$a_n = * \frac{1}{\zeta(k)} \sum_{d|n} d^{k-1} = \frac{*}{\zeta(k)} \, \sigma_{k-1}(n), \tag{1.4.10}$$

where $*$ is independent of n. The general case is not much more complicated. It is always a sum of divisors of n, i.e., of the type

$$\sum_{d|n} F(d) \qquad \text{where } F \text{ is periodic (mod } N). \tag{1.4.11}$$

Remarks 1.4.4.

(A) By an indentical calculation one can develop the spectral Eisenstein series (1.4.3) in a Fourier series. For example if $\Gamma = \Gamma(1)$ then

$$E(z,s) = y^s + \frac{\zeta^*(2s-1)}{\zeta^*(2s)} y^{1-s}$$

$$+ \frac{2y^{1/2}}{\zeta^*(2s)} \sum_{n=1}^{\infty} n^{s-1/2} \, \sigma_{1-2s}(n) \, K_{s-1/2}(2\pi n y) \, \cos(2\pi n x) \tag{1.4.12}$$

where

$$\zeta^*(s) = \pi^{-s/2} \Gamma(s/2) \zeta(s)$$

$$K_\nu(t) = \int_0^\infty e^{-t \cosh u} \cosh(\nu u) du.$$

From this representation and the well known analytic continuation of the Riemann zeta function, it is clear that $E(z,s)$ has a meromorphic continuation to the plane. We will use (1.4.12) in Appendix 1.1 to show that $\Delta(z)$ is modular.

(B) The analysis of $E_k(z)$ above works for $k > 2$. For $k = 2$ we need to modify the construction to overcome the convergence difficulty. Introduce for $s > 0$

$$E_k(z,s) = \sum_{\gamma \in \Gamma_\infty \backslash \Gamma} (cz+d)^{-2} |cz+d|^{-2s}.$$

One can show that $\lim_{s \to 0} E_2(z,s) = E_2(z,0)$ exists and yields a function which transforms correctly but which is not quite holomorphic. In fact its zeroth coefficients have a simple non holomorphic part. Similarly one constructs $E_2^{(j)}(z,0)$ for each cusp p_j, $j = 1, \ldots, r$, of Γ. Any combination of these $E_2^{(j)}(z,0)$ annihilating the non holomorphic term gives an element of $\mathcal{M}_2(\Gamma)$. In this way one gets an $r-1$ dimensional space $\mathcal{E}_2(\Gamma)$ of $\mathcal{M}_2(\Gamma)$. The Fourier coefficients of a member e of $\mathcal{E}_2(\Gamma)$ are again divisor sums. Note also that an arbitrary $g \in \mathcal{M}_2(\Gamma)$ gives a meromorphic differential $g(z)dz$ over $\Gamma \backslash \mathbf{H}$ (compactified). Since the sum of the residues of $g(z)dz$ over $\Gamma \backslash \mathbf{H}$ is zero it follows that the values of $g(z)$ at $p_1, \ldots, p_r$ satisfy a linear relation. So everything works out as before. The Eisenstein series span an $r-1$ dimensional subspace $\mathcal{E}_2(\Gamma)$ of $\mathcal{M}_2(\Gamma)$ and every $g \in \mathcal{M}_2(\Gamma)$ is uniquely expressible as

$$g = e + h \tag{1.4.13}$$

with $e \in \mathcal{E}_2(\Gamma)$, $h \in S_2(\Gamma)$. For more details concerning Eisenstein series of weight 2, see [Sc2].

(C) We can now explain Ramanujan's observation (1.1.4). We have seen in (1.3.11) that for $s \geq 4$

$$\theta_s(z) = \sum_{\nu=0}^{\infty} r_s(n)e(nz) \in \mathcal{M}_{s/2}(\Gamma_0(4)).$$

Hence from the considerations of Eisenstein series

$$\theta_s(z) = e_s(z) + h_s(z) \tag{1.4.14}$$

where $e_s \in \mathcal{E}_{s/2}(\Gamma_0(4))$ is an Eisenstein series, $h_s \in S_{s/2}(\Gamma_0(4))$.

Equating Fourier coefficients gives

$$r_s(n) = \delta_s(n) + h_s(n) \tag{1.4.15}$$

where $\delta_s(n)$ is the coefficient of some Eisenstein series and is therefore a divisor sum (or singular series as it is called in the 'circle method' [Da2]) while $h_s(n)$ by the Ramanujan conjecture on cusp forms should satisfy

$$h_s(n) = O_\epsilon(n^{\frac{s}{4}-\frac{1}{2}+\epsilon}).$$

For small values of s, say $s = 4$, there are no cusp forms, i.e., $S_2(\Gamma_0(4)) = \{0\}$. For this case, $s = 4$, $\Gamma_0(4)$ (compactified) is easily seen to be of genus zero so that the space of holomorphic Abelian differentials $f(z)dz$ on $\Gamma_0(4)\backslash\mathbf{H}$ is zero dimensional. A cusp form $f \in S_2(\Gamma_0(4))$ gives rise to such a differential, hence $S_2(\Gamma_0(4)) = \{0\}$. Consequently

$$r_4(n) = \delta_4(n).$$

A straightforward calculation with Eisenstein series gives

$$r_4(n) = 8 \sum_{\substack{d|n \\ 4\nmid d}} d,$$

viz. Jacobi's result (1.1.6).

1.5   Poincaré series

As we have seen, Eisenstein series never furnish cusp forms. To obtain cusp forms by a similar construction we use a variation of Eisenstein series known as Poincaré series. Consider the group $\Gamma_0(N)$ (for $\Gamma(N)$ the analysis is the same) where, whenever we are dealing with $1/2$ odd integer weight, $4|N$. The Poincaré series at '∞', $P_m(z,k)$ is defined by

$$P_m(z,k) = \sum_{\gamma \in \Gamma_\infty \backslash \Gamma} (j(\gamma,z))^{-2k} \, e(m\gamma z). \qquad (1.5.1)$$

Here $m \geq 0$ is an integer. Actually $m = 0$ is just the Eisenstein series so for this section we assume that $m > 0$. Clearly the Poincaré series are dominated by the Eisenstein series and so for $k > 2$, which we assume here for simplicity, the above series converges absolutely. As before $P_m(z,k)$ vanishes at all cusps not equivalent to ∞. However, it also vanishes at ∞ since $m > 0$. Hence $P_m(z,k) \in S_k(\Gamma)$ for each $m \geq 1$ (it may well vanish identically). The $P_m(z,k)$, $m \geq 1$ span the space of cusp forms $S_k(\Gamma)$ as we now demonstrate.

1.5.1. Petersson inner product.

So far we have viewed $S_k(\Gamma)$ only as a linear space. It can be given a natural inner product called the *Petersson inner product*. For $f, g \in S_k(\Gamma)$

$$\langle f, g \rangle = \int_{\Gamma\backslash\mathbf{H}} y^k f(z) \overline{g(z)} \, \frac{dxdy}{y^2}. \qquad (1.5.2)$$

The function $y^k f(z) \overline{g(z)}$ is Γ–invariant and rapidly decreasing in the cusps, hence the integral is well defined and convergent. The reason that this inner product is natural is that all interesting linear operators on $S_k(\Gamma)$ (see for example the Hecke operators later on) are Hermitian (or at least normal) relative to this inner product.

The following is a basic calculation: Let $f \in S_k(\Gamma)$ then

$$\langle P_m, f \rangle = \int_{\Gamma \backslash \mathbf{H}} P_m(z) \overline{f(z)} y^k \frac{dx\,dy}{y^2}$$

$$= \int_{\Gamma \backslash \mathbf{H}} \sum_{\gamma \in \Gamma_\infty \backslash \Gamma} (j(\gamma, z))^{-2k} e(m\gamma z) \overline{f(z)} y^k \frac{dx\,dy}{y^2} \quad (1.5.3)$$

$$= \int_0^\infty \int_0^1 e(mz) \overline{f(z)} y^k \frac{dx\,dy}{y^2}$$

$$= \frac{\overline{a_m}}{(4\pi m)^{k-1}} \Gamma(k-1),$$

where $f(z) = \sum_{n=1}^\infty a_n e(nz)$ and Γ is the gamma function. It follows that if $\langle P_m, f \rangle = 0$ for all m then $a_m = 0$ for all m, i.e., $f \equiv 0$. Hence the P_m's span the finite dimensional space $S_k(\Gamma)$.

It follows that for our main purpose of estimating Fourier coefficients of cusp forms it suffices to do so for the Poincaré series P_m. Let us see what this involves. We can compute the Fourier development of these Poincaré series, the calculation being very similar to the one involving Eisenstein series.

$$\int_0^1 P_m(z, k) e(-nz)\,dz$$

$$= \int_0^1 \sum_{\gamma \in \Gamma_\infty \backslash \Gamma} (j(\gamma, z))^{-2k} e(m\,\gamma\,z) e(-nz)\,dz$$

$$= 2 \int_0^1 e(mz)\, e(-nz)\,dz + \sum_{\substack{c \neq 0 \\ \gamma = \left(\begin{smallmatrix} a & b \\ c & d \end{smallmatrix}\right) \in \Gamma_\infty \backslash \Gamma / \Gamma_\infty}}$$

$$\sum_r \int_0^1 j\left(\gamma \left(\begin{smallmatrix} 1 & r \\ 0 & 1 \end{smallmatrix}\right), z\right)^{-2k} e\left(m\gamma \left(\begin{smallmatrix} 1 & r \\ 0 & 1 \end{smallmatrix}\right) z - nz\right)\,dz$$

$$= 2\delta_{m,n} + \sum_{\substack{c \neq 0 \\ \gamma \in \Gamma_\infty \backslash \Gamma / \Gamma_\infty}} \int_{-\infty}^\infty j(\gamma, z)^{-2k} e(m\gamma z - nz)\,dz$$

Now $\gamma z = \dfrac{a}{c} - \dfrac{1}{c(cz+d)}$ so

$$= 2\delta_{m,n} + 2 \sum_{\substack{c>0 \\ \gamma \in \Gamma_\infty \backslash \Gamma / \Gamma_\infty}}$$

$$\sum_{\substack{d \,(\mathrm{mod}\, c) \\ (d,c)=1}} \varepsilon_d^{-2k} \left(\frac{c}{d}\right)^{2k} \int_{-\infty}^{\infty} (cz+d)^{-k} e\left(\frac{ma}{c} - \frac{m}{c(cz+d)} - nz\right) dz$$

$$= 2\delta_{m,n} + 2 \sum_{\substack{c>0 \\ c\equiv 0 \,(\mathrm{mod}\, N)}} c^{-k}$$

$$\sum_{d \,(\mathrm{mod}\, c)} e\left(\frac{ma}{c}\right) \varepsilon_d^{-2k} \left(\frac{c}{d}\right)^{2k} \int_{-\infty}^{\infty} \left(z+\frac{d}{c}\right)^{-k} e\left(\frac{-m}{c^2(z+d/c)} - nz\right) dz$$

$$= 2\delta_{m,n} + 2 \sum_{\substack{c\equiv 0 \,(\mathrm{mod}\, N) \\ c>0}} c^{-k}$$

$$\sum_{\substack{d \,(\mathrm{mod}\, c) \\ (d,c)=1}} \left(\frac{c}{d}\right)^{2k} \varepsilon_d^{-2k} e\left(\frac{ma}{c}\right) \int_{-\infty}^{\infty} z^{-k} e\left(\frac{-m}{c^2 z} - n\left(z - \frac{d}{c}\right)\right) dz$$

$$= 2\delta_{m,n} + 2 \sum_{\substack{c\equiv 0 \,(\mathrm{mod}\, N) \\ c>0}} c^{-k}$$

$$\left(\sum_{\substack{d \,(\mathrm{mod}\, c) \\ (d,c)=1}} \left(\frac{c}{d}\right)^{2k} \varepsilon_d^{-2k} e\left(\frac{ma+nd}{c}\right) \int_{-\infty}^{\infty} z^{-k} e\left(\frac{-m}{c^2 z} - nz\right) dz \right)$$

So

$$\hat{P}_m(n) = 2\left(\frac{m}{n}\right)^{(k-1)/2}$$

$$\left\{ \delta_{m,n} + 2\pi i^{-k} \sum_{\substack{c\equiv 0 \,(\mathrm{mod}\, N) \\ c>0}} J_{k-1}\left(\frac{4\pi\sqrt{mn}}{c}\right) \frac{K(m,n,c)}{c} \right\}, \quad (1.5.4)$$

where we have used the representation for the Bessel function

$$\int_{-\infty+ci}^{\infty+ci} w^{-k} \exp(-(\mu_1 w + \mu_2 w^{-1})) dw$$

$$= 2\pi \left(\frac{\mu_1}{\mu_2}\right)^{(k-1)/2} e^{-ik\pi/2} J_{k-1}(4\pi\sqrt{\mu_1\mu_2}),$$

while $P_m(z, k) = \sum_{n=1}^{\infty} \hat{P}_m(n)\, e(nz)$ and

$$K(m, n, c) = \sum_{\substack{d \pmod c \\ (d,c)=1}} \left(\frac{c}{d}\right)^{2k} \varepsilon_d^{-2k} e\left(\frac{m\bar{d} + nd}{c}\right). \qquad (1.5.5)$$

((1.5.4) is of course an algebraist's nightmare; one expresses a good integer like $\tau(n)$ as an infinite series with Bessel functions!)

To continue we must distinguish between the fundamentally different sums in (1.5.5) when k is even and when k is $1/2$ an odd integer.

1.5.2. k even.

In this case the exponential sum $K(m, n, c)$ above is a so-called Kloosterman sum

$$K(m, n, c) = K(n, m, c) = \sum_{\substack{d \pmod c \\ (d,c)=1}} e\left(\frac{m\bar{d} + nd}{c}\right). \qquad (1.5.6)$$

The Kloosterman sums satisfy an obvious property which follows from the Chinese remainder theorem. If $(c_1, c_2) = 1$ then

$$K(u, v, c_1 c_2) = K(u, v\bar{c}_2^2, c_1)\, K(u, v\bar{c}_1^2, c_2), \qquad (1.5.6')$$

$K(u, v, p^k)$ for $k \geq 2$ is easily (elementarily) seen to satisfy

$$|K(u, v, p^k)| \leq 2 p^{k/2}. \qquad (1.5.7)$$

If we combine this with the important Weil bound [We1], that is for p prime and $(m, n, p) = 1$;

$$\left| \sum_{x \pmod p} e\left(\frac{mx + n\bar{x}}{p}\right) \right| \leq 2\sqrt{p}, \qquad (1.5.8)$$

we see that for m fixed

$$K(m, n, c) = O_\epsilon(c^{1/2+\epsilon}). \qquad (1.5.9)$$

Using this bound in the series (1.5.4) together with the bound for the Bessel function [Wa]

$$J_{k-1}(x) \ll \min\left\{ x^{k-1}, \frac{1}{\sqrt{x}} \right\} \qquad (1.5.10)$$

we conclude

$$\hat{P}_m(n) = O_\epsilon(n^{k/2-1/4+\epsilon}). \qquad (1.5.11)$$

We have proven the following (after appropriate extensions to $\Gamma(N)$ and $k = 2$).

Proposition 1.5.3. *Let k be an even positive integer, if $f \in S_k(\Gamma)$ with Γ a congruence subgroup then the Fourier coefficients a_n of f satisfy*

$$a_n = O_\epsilon(n^{k/2-1/4+\epsilon}).$$

This is a first and non-trivial step toward the Ramanujan conjectures 1.2.5. Though it falls short it suffices for many applications and in more generality such as automorphic forms over a number field it yields the best known results; see the notes at the end of Chapter 1.

1.5.4. k half an odd integer.

This time

$$K(m,n,c) = \sum_{d \,(\mathrm{mod}\, c)} \left(\frac{c}{d}\right) \varepsilon_d^{-1} e\left(\frac{m\bar{d}+nd}{c}\right). \qquad (1.5.12)$$

This sum is essentially a Salié sum, see Chapter 4. We will study it in detail there. Here, however, we merely note that again it factorizes and the estimate

$$|K(m,n,p)| \le 2\sqrt{p}, \qquad \text{for } p \text{ a prime}$$

is in this case elementary (see Chapter 4). We conclude, as before

Proposition 1.5.5. *Let k be half an odd integer and $f \in S_k(\Gamma_0(N))$, $(4|N)$, then*

$$a_n = O(n^{k/2-1/4+\epsilon}).$$

In view of Example (1.3.8) we see that this is, in fact, the best estimate one can claim here without restricting to n to be square free or avoiding the θ–series of one variable.

Clearly if one is to improve on these bounds one needs to exploit the cancellation that occurs in the series (1.5.4) that comes form the sign changes in $K(m,n,c)$. This is a difficult problem. In this direction Linnik [Li2] and Selberg [Se] have conjectured

Conjecture 1.5.6. (Linnik–Selberg)

$$\sum_{0<c\le x} \frac{K(m,n,c)}{c} = O(x^\epsilon) \qquad \text{for } x > (m,n)^{1/2+\epsilon}.$$

This would imply Conjectures 1.2.5 and some progress in the direction of Conjecture 1.5.6 is described in Appendix 1.2. The truth of 1.5.6 has many other implications in number theory, see Linnik [Li2].

The exploitation of this cancellation for the Salié sums is the subject matter of Chapter 4.

Returning to Remark 1.4.4(c) we see that for $s \geq 3$;

$$r_s(n) = \delta_s(n) + O_\epsilon(n^{s/4 - 1/4 + \epsilon}). \qquad (1.5.13)$$

The singular series $\delta_s(n)$ is of order $n^{s/2-1}$ (when $s = 3$ this is so only for those n's representable as a sum of three squares) and so for $s > 3$, $\delta_s(n)$ is the main term. For $s = 3$ it is not, and this is the reason the Linnik problem (C) of the Introduction is a difficult one. One needs to go beyond the $k/2 - 1/4$ barrier!

1.6 Hecke operators

We have discussed in some detail the second observation of Ramanujan viz. (1.1.2). The first is explained by '*Hecke operators*'.

Quite generally if G is a group and $\Gamma \leq G$ such that G acts on a space X with Γ acting discontinuously on X, we can define certain Hecke operators on $L^2(\Gamma \backslash X)$. Let

$$\text{COM}\,(\Gamma) \overset{\text{def}}{=} \{g \in G |\ \Delta = g^{-1}\Gamma g \cap \Gamma \text{ has finite index in both } \Gamma \text{ and } g^{-1}\Gamma g\}.$$

Clearly COM (Γ), called the commensurator of Γ, is a subgroup of G containing Γ. For each $g \in \text{COM}\,(\Gamma)$ define

$$T_g : L^2(\Gamma \backslash X) \to L^2(\Gamma \backslash X)$$

as follows: Write

$$\Gamma = \bigcup_{j \in F} \Delta \delta_j,$$

i.e., in right cosets, then

$$F(x) = T_g f(x) = \sum_{j \in F} f(g \delta_j x). \qquad (1.6.1)$$

Claim: $F(x) \in L^2(\Gamma \backslash X)$.

Indeed

$$F(\gamma x) = \sum_{j \in F} f(g \delta_j \gamma x),$$

now $\delta_j\gamma = \delta_j^1\delta_{\pi(j)}$ for some permutation π of F. Hence

$$F(\gamma x) = \sum_j f(g\delta_j^1\delta_{\pi(j)}x) = \sum_j f(g\,g^{-1}\mu_j g\delta_{\pi(j)}x)\,,$$

for some $\mu_j \in \Gamma$. Thus

$$F(\gamma x) = \sum_j f(g\,\delta_j x) = F(x)\,,$$

proving the claim.

Of course T_g is a trivial operator if $g \in \Gamma$, but otherwise if $\mathrm{COM}(\Gamma) \supsetneq \Gamma$, T_g and especially its spectrum is interesting. The set of such Hecke operators form an algebra which, for example, will commute with all G invariant differential operators on X. In order that the spaces of certain functions on $\Gamma\backslash X$ be finite dimensional, we usually need $\Gamma\backslash X$ to be compact or at worst of finite volume. If G is compact then we can take $\Gamma = \{\mathrm{id.}\}$ and every $g \in G$ is a commensurator. The elements of the algebra of such Hecke operators will be of importance in Chapter 2 (primarily with $X = S^2$, $G = SO(3)$).

When $G = SL(2,\mathbf{R})$ and $\Gamma = \Gamma(N)$, we obtain from the above construction the usual Hecke operators. We consider here only $\Gamma = \Gamma(1)$ and will be brief since no direct use of these operators will be made.

For $\Gamma(1)$, $G = GL(2,\mathbf{R})$, we have $g = \binom{n\,0}{0\,1} \in \mathrm{COM}(\Gamma(1))$. A simple calculation shows that the corresponding Hecke operators on $L^2(\Gamma\backslash SL(2,\mathbf{R}))$ are

$$T_{\binom{n\,0}{0\,1}}f(h) \triangleq T_n f(h) = \sum_{\substack{m\geq 1\\mr=n\\k\,(\mathrm{mod}\,r)}} f\left(\begin{pmatrix} m & k \\ 0 & r \end{pmatrix}h\right)\,.$$

Now one checks that a function f on $\mathbf{H}$ satisfies

$$f(\gamma z) = (cz+d)^k f(z) \qquad (k \text{ even integer})$$

iff $F = y^{k/2} f(z)$, thought of on $SL(2,\mathbf{R})$, satisfies

$$F\left(g\begin{pmatrix} \cos\theta & \sin\theta \\ -\sin\theta & \cos\theta \end{pmatrix}\right) = e(k\theta) F(g) \qquad (i)$$

$$F(\gamma g) = F(g) \qquad (ii)$$

Here we identify $\mathbf{H}$ with $SL(2,\mathbf{R})/K$,

$$K = \left\{\begin{pmatrix} \cos\theta & \sin\theta \\ -\sin\theta & \cos\theta \end{pmatrix}\,\middle|\, \theta \in \mathbf{R}\right\}\,,$$

and

$$g = \begin{pmatrix} 1 & x \\ 0 & 1 \end{pmatrix} \begin{pmatrix} y^{1/2} & 0 \\ 0 & y^{-1/2} \end{pmatrix} \begin{pmatrix} \cos\theta & \sin\theta \\ -\sin\theta & \cos\theta \end{pmatrix}$$

is the usual Iwasawa factorization of g and $z = x + iy$.

Carrying out the identification we find that for functions on **H** satisfying

$$f(\gamma z) = (cz + d)^k f(z)$$

the action of T_n becomes

$$T_n f(z) = n^{k-1} \sum_{\substack{ad=n,\, a\geq 1 \\ b \,(\mathrm{mod}\, d)}} d^{-k} f\left(\frac{az+b}{d}\right). \qquad (1.6.2)$$

One can check that

(i) $\qquad\qquad T_{nm} = T_n T_m \qquad$ for $(n,m) = 1$

(ii) $\qquad T_{p^n} T_p = T_{p^{n+1}} + p^{n-1} T_{p^{n-1}}, \qquad n \geq 1$

(iii) $\qquad\qquad f \in \mathcal{M}_k(\Gamma(1))$ implies $T_n f \in \mathcal{M}_k(\Gamma(1))$

Moreover if

$$f(z) = \sum_{n=0}^{\infty} a_n e(nz)$$

then

$$T_n f(z) = \sum_{m=0}^{\infty} b_m e(mz)$$

with

$$b_m = \sum_{d|(n,m)} d^{k-1} a\left(\frac{mn}{d^2}\right).$$

In particular, if $T_n f = \lambda_n f$ then

(iv) $\qquad\qquad\qquad\qquad \lambda_n a_1 = a(n).$

It follows from (iv) that if f is an eigenfunction of all the Hecke operators T_n then the Fourier coefficients of f inherit properties (i) and (ii).

Corollary 1.6.1. *The coefficients $\tau(n)$ of $\Delta(z)$ are multiplicative (i.e., $\tau(mn) = \tau(m)\tau(n)$ if $(n,m) = 1$) and*

$$\tau(p)\tau(p^n) = \tau(p^{n+1}) + p^{11}\tau(p^{n-1}), \qquad n \geq 1.$$

Proof: The result will certainly follow if $S_{12}(\Gamma(1))$ is one dimensional since, as is shown in Appendix 1.1, $\Delta(z) \in S_{12}(\Gamma(1))$. This is because $T_n : S_{12}(\Gamma(1)) \to S_{12}(\Gamma(1))$, and Δ would then have to be an eigenfunction of all Hecke operators. To see that $S_{12}(\Gamma(1))$ is one dimensional consider

$$\frac{1}{2\pi i} \int_{\partial \mathcal{F}_{\Gamma(1)}} (f'/f) dz \,.$$

One finds by direct calculation that this equals 1. On the other hand it also equals the number of zeroes of f. If $\dim S_{12}(\Gamma(1)) \geq 2$ then there would be a nonzero $f \in S_{12}(\Gamma(1))$ vanishing to order at least 2 at ∞ which would be impossible in view of the above.   □

We can now also explain the more precise bound

$$|\tau(n)| \leq d(n)\, n^{11/2}$$

conjectured by Ramanujan. From Corollary 1.6.1 this would follow from $|\tau(p)| \leq 2 p^{11/2}$. Indeed from Corollary 1.6.1

$$\tau(p^k) = p^{11\,k/2} \frac{\sin((k+1)\theta)}{\sin(\theta)}$$

where

$$\cos \theta = \frac{\tau(p)}{2 p^{11/2}} \,.$$

Hence $\tau(n) = O(n^{11/2+\epsilon})$ iff $|\tau(p)| \leq 2 p^{11/2}$ for primes p.

As was mentioned we have been very brief in Section 1.6. For a discussion of Hecke operators see Ogg [O] or Serre [Ser].

Appendix 1.1

In this Appendix we prove that $\Delta(z) = \sum_{n=1}^{\infty} \tau(n) e(nz)$ is a cusp form of weight 12 for $\Gamma(1)$. More in fact is shown. Let

$$\eta(z) = e^{i\pi z/12} \prod_{n=1}^{\infty} (1 - e(nz)) \,. \tag{A.1.1}$$

We show that $\eta(z)$ is a modular form of $1/2$ integral weight.

Proposition A.1.1. *The function* $F(z) = y^{1/2}|\eta(z)|^2$ *is* $\Gamma(1)$ *invariant.*

Once Proposition A.1.1 is established we will have

(i) $$\eta(z+1) = e\left(\frac{1}{24}\right) \eta(z) \,.$$

Secondly $|\eta(-1/z)| = |z|^{1/2} |\eta(z)|$ hence

$$\eta(-1/z) = c\,\eta(z)\,z^{1/2}$$

with $|c| = 1$. Evaluating this at $z = i$ gives

$$\eta(i) = c\,e^{i\pi/4}\,\eta(i)$$

and since $\eta(z) \neq 0$ for $z \in \mathbf{H}$ as is clear from its definition in (A.1.1), we conclude

(ii) $$\eta(-1/z) = e^{-i\pi/4}z^{1/2}\eta(z)\,.$$

From (i) and (ii) it follows that

$$\Delta(z) = (\eta(z))^{24}$$

satisfies

(i)′ $$\Delta(z+1) = \Delta(z)$$

(ii)′ $$\Delta(-1/z) = z^{12}\,\Delta(z)\,.$$

Since $\binom{1\ 1}{0\ 1}$ and $\binom{0\ \ 1}{-1\ 0}$ generate $\Gamma(1)$, the modularity of Δ follows. $\qquad\square$

To prove Proposition A.1.1 we will interpret the quantity $F(z)$ in a way which shows that it depends on the torus $L\backslash\mathbf{C}$, where $L = \{m + nz\,|\,m, n \in \mathbf{Z}\}$, and *not* on a basis for L.

Consider the series

$$E^*(z, s) = {\sum_{m,n}}' \frac{y^s}{|mz + n|^{2s}}\,, \qquad \text{where } {\sum}' \text{ means omit } (m, n) = (0, 0)\,.$$

$$\text{(A.1.2)}$$

This is a close relative of the spectral Eisenstein series (1.4.3). In fact from the coset representatives

$$\Gamma_\infty\backslash\Gamma(1) = \left\{ \begin{pmatrix} * & * \\ c & d \end{pmatrix} \,\middle|\, (c, d) = 1 \right\} \qquad \text{(A.1.3)}$$

$E^*(z, s)$ has a nice geometric interpretation. Let M be the flat torus $L\backslash\mathbf{R}^2$, $L = \{m + nz\,|\,m, n \in \mathbf{Z}\}$. The spectrum of Δ, the Laplacian on functions on M, is easily determined. The eigenfunctions of Δ are

$$\phi(x) = e(\langle \ell^*, x\rangle)$$

where $x = (x_1, x_2)$, $\ell^* \in L^* = \{y \in \mathbf{R}^2\,|\, \langle y, \ell\rangle \in \mathbf{Z} \text{ for all } \ell \in L\}$. The corresponding eigenvalue is

$$-4\pi^2\,|\ell^*|^2\,, \qquad \ell^* \in L^*\,.$$

The spectral zeta function $\zeta_M(s)$ for M is defined as follows

$$\zeta_M(s) = \sum{}'(-\lambda)^{-s} = \sum_{\ell^* \in L^*}{}'(4\pi^2 |\ell|^2)^{-s}. \qquad (A.1.4)$$

We see that $\zeta_M(s)$ is essentially the Eisenstein series $E^*(z,s)$. The quantity $\zeta_M'(0)$ is formally

$$\zeta_M'(0) = -\sum{}' \log \lambda$$

hence

$$e^{-\zeta_M'(0)} = \det{}' \Delta_z, \qquad (A.1.5)$$

the 'regularized determinant' of Δ. Our proof of Proposition A.1.1 is to compute $\det' \Delta_z$ explicitly and in particular its Fourier development.

We have that $E^*(z,s)$ is a $\Gamma(1)$ invariant function in z. Thus the 'height' function $h(z) = -\log \det' \Delta_z$ given by

$$h(z) = \frac{\partial}{\partial s} E^*(z,s) \Big|_{s=0} \qquad (A.1.6)$$

is clearly a modular function for $\Gamma(1)$. To compute the derivative at $s = 0$ we use the expansion (1.4.9)

$$E^*(z,s) = \zeta(2s) y^s + \frac{\zeta^*(2s-1)\pi^s}{\Gamma(s)} y^{1-s}$$
$$+ \frac{4y^{1/2}\pi^s}{\Gamma(s)} \sum_{n=1}^{\infty} n^{s-1/2} \sigma_{1-2s}(n) K_{s-1/2}(2\pi ny) \cos(2\pi nx).$$

Hence

$$\frac{\partial E^*}{\partial s}\Big|_{s=0} = \zeta(0) \log y + 2\zeta'(0) + \zeta^*(-1)y + 4y^{1/2}$$
$$\sum_{n=1}^{\infty} n^{-1/2} \sigma_1(n) K_{-1/2}(2\pi ny) \cos(2\pi nx).$$

One can evaluate $K_{-1/2}(y)$ as a more elementary function (this is an exercise but its importance in Chapter 4, see 4.10, cannot be over emphasized)

$$K_{-1/2}(y) = \sqrt{\frac{\pi}{2y}} e^{-y}$$

so that the above becomes

$$\frac{\partial E^*}{\partial s}\Big|_{s=0} = \zeta(0) \log y + 2\zeta'(0) + \zeta^*(-1)y + 2\sum_{n=1}^{\infty} n^{-1} \sigma_1(n) e^{-2\pi ny} \cos(2\pi nx).$$

Now, as is well known [GR], $\zeta(0) = -1/2$, $\zeta'(0) = -\frac{1}{2}\log(2\pi)$ and $\zeta^*(1-s) = \zeta^*(s)$ so

$$\left.\frac{\partial E^*}{\partial s}\right|_{s=0} = \text{Re}\left\{-\frac{1}{2}\log y - \log(2\pi) + \zeta^*(2)y + 2\sum_{n=1}^{\infty} n^{-1}\sum_{d|n} d\,e(nz)\right\}.$$

Also $\zeta^*(2) = \pi^{-1}\Gamma(1)\zeta(2) = \frac{\pi}{6}$, hence

$$-\left.\frac{\partial E^*}{\partial s}\right|_{s=0} = \text{Re}\left\{\log\left[2\pi y^{1/2}e^{i\pi/6}\prod_{m=1}^{\infty}(1 - e(mz))^2\right]\right\},$$

i.e.,

$$-h(z) = \log(2\pi y^{1/2}|\eta(z)|^2). \tag{A.1.7}$$

We have seen that $h(z)$ is modular and so the Proposition is proven. □

In fact we have shown that

$$\det{}'\Delta_z = C\,y^{1/2}|\eta(z)|^2, \tag{A.1.8}$$

where C is a constant independent of z.

This result has been rediscovered by physicists in the context of string theory on a genus 1 Riemann surface [Pol]. The above proof of (A.1.8) goes through the evaluation of $-\frac{\partial}{\partial s}E^*(z,s)\big|_{s=0}$ in terms of $\eta(z)$, which is a result due to Kronecker [Kr] known as Kronecker's first limit formula.

Appendix 1.2.

In this Appendix we show how the theory of automorphic forms may be used to obtain some progress towards the Linnik–Selberg Conjecture 1.5.6. As was pointed out originally by Linnik [Li2], Conjecture 1.5.6 has many applications to number theory (besides the Ramanujan conjectures) and in fact this circle of ideas has led through the work of Deshouillers and Iwaniec [DI] to some striking applications, see [Iw1]. Another application of cancellation of 'Kloosterman sums' was found by Vardi [V], see notes to this Chapter.

Theorem A.2.1.
$$\sum_{c\leq x}\frac{K(m,n,c)}{c} = O_\epsilon(x^{1/6+\epsilon}).$$

The direct use of Weil's bound (1.5.8) leads to $O_\epsilon(x^{1/2+\epsilon})$, so that Theorem A.2.1 represents an exploitation of the cancellations due to the signs of $K(m,n,c)$ (actually this seems most certainly a true statement though, as pointed out by Serre, it is not clear that $K(m,n,c) = O(c^{1/2-\epsilon_0})$ for $\epsilon_0 > 0$ is impossible for m,n fixed, $c\to\infty$).

To prove Theorem A.2.1 we need to consider more general automorphic forms on $\mathbf{H}$ than the ones that we have considered so far – these come about by considering eigenfunctions of invariant differential operators.

Let $\Delta = y^2 \left(\frac{\partial^2}{\partial x^2} + \frac{\partial^2}{\partial y^2} \right)$ acting on functions on $\mathbf{H}$. Δ is the Laplacian for the hyperbolic metric $ds = |dz|/y$ on $\mathbf{H}$. Since $G = SL(2, \mathbf{R})$ acting on $\mathbf{H}$ by the usual linear fractional action acts as isometries, we first find that Δ commutes with G. Hence we may consider its action on Γ invariant functions, where $\Gamma < G$ is a congruence group. The Laplacian Δ acting on the space $L^2(\Gamma \backslash \mathbf{H})$ is self adjoint and we may investigate its spectrum (the inner product on $\Gamma \backslash \mathbf{H}$ is the usual Petersson inner product $\langle f, g \rangle = \int_{\Gamma \backslash \mathbf{H}} f(z) \overline{g(z)} \, dxdy/y^2$).

For example the function y^s on $\mathbf{H}$ satisfies

$$\Delta(y^s) + s(1 - s) \, y^s = 0 \,. \tag{A.2.1}$$

Hence the series (A.1.2) $E^*(z, s)$ correspondingly satisfies

$$\Delta E^*(z, s) + s(1 - s) \, E^*(z, s) = 0 \,,$$

i.e., the $E^*(z, s)$ are $\Gamma(1)$ invariant eigenfunctions of Δ. From the Fourier expansion (1.4.9) it is apparent (by looking at the 'constant term', i.e., $n = 0$) that they are not in $L^2(\Gamma(1) \backslash \mathbf{H})$. Though this will not concern us here we remark that these $E^*(z, s)$ may be used to furnish the 'continuous spectrum' of $\Gamma(1) \backslash \mathbf{H}$.

Returning to $L^2(\Gamma \backslash \mathbf{H})$ we note that for ϕ in the domain of Δ, we have

$$-\int_{\Gamma \backslash \mathbf{H}} \Delta \phi \cdot \overline{\phi} \, \frac{dxdy}{y^2} = \int_{\Gamma \backslash \mathbf{H}} |\nabla \phi|^2 dxdy \geq 0 \,.$$

Hence $\lambda_0 = 0$ is the smallest eigenvalue of Δ on $\Gamma \backslash \mathbf{H}$ corresponding to the constant function. We let

$$\lambda_1 = \inf_{\substack{\phi \in \mathrm{dom}\,(\Delta) \\ \int_{\Gamma \backslash \mathbf{H}} \phi \frac{dxdy}{y^2} = 0}} \frac{\int_{\Gamma \backslash \mathbf{H}} |\nabla \phi|^2 dxdy}{\int_{\Gamma \backslash \mathbf{H}} |\phi|^2 dxdy} \tag{A.2.2}$$

be the next smallest eigenvalue (actually if $\Gamma \backslash \mathbf{H}$ is not compact, as we are assuming, λ_1 need not be an 'eigenvalue' but it is the next point in the spectrum of Δ).

Proposition A.2.2.

$$\lambda_1(\Gamma(1) \backslash \mathbf{H}) \geq \frac{1}{4} \,.$$

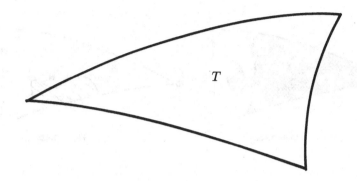

Figure 1.3: Triangle $T \subset \mathbf{H}$

In fact Selberg [Se] has conjectured

$$\lambda_1(\Gamma(N)\backslash\mathbf{H}) \geq 1/4 \qquad \text{for all } N, \tag{A.2.3}$$

and (A.2.3) may be viewed as a certain generalization of the Ramanujan conjectures. We will return to this a little later.

To prove A.2.2, consider the general problem of bounding $\lambda_1^{(N)}(\Omega)$ where $\Omega \subset \mathbf{H}$ is a nice domain in $\mathbf{H}$ and $\lambda_1^{(N)}$ indicates the next-to-largest eigenvalue of Δ on Ω with Neumann (or free) boundary conditions.

Proposition A.2.3. *Let T be a triangle (hyperbolic of finite area) in $\mathbf{H}$, then $\lambda_1^{(N)}(T) \geq 1/4$.*

Proof: We prove this for compact triangles (see Figure 1.3). The proof for a triangle with an angle of zero at some vertex is similar (though technically a little more involved). We have $\lambda_0^N(T) = 0$, corresponding to $f_0(z) = \text{const}$. Let $\lambda_1^N(T) = \lambda_1$ be the next largest eigenvalue with corresponding eigenfunction $\phi_1(z)$. We have $\int_T \phi_1(z)d\mu(z) = 0$. Hence $\phi_1(z)$ has a nodal set (i.e., the set where it vanishes) $\mathcal{N}$, part of which must either run from one edge to another as shown in Figure 1.4, or it looks like Figure 1.5. Note that by reflecting Ω in Figure 1.4.b about OA we get a region like the Ω of Figure 1.5.

Now consider the domain Ω shown. On integrating by parts we have

$$\lambda_1 = \frac{\displaystyle\int_\Omega |\nabla_H \phi|^2 d\mu(z)}{\displaystyle\int_\Omega |\phi|^2 d\mu(z)}.$$

We use geodesic polar coordinates (r, θ) in the Ω of Figure 1.4, from O (the case 1.5 is a special case with angle $\alpha = 2\pi$). In these coordinates

$$ds^2 = dr^2 + (\sinh r)^2 d\theta^2.$$

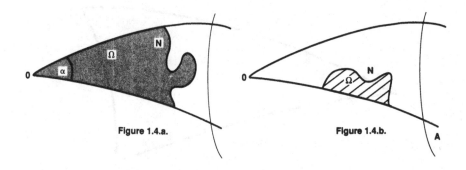

Figure 1.4: Nodal sets

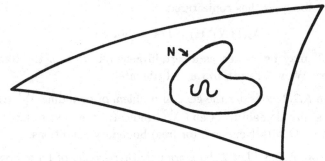

Figure 1.5: Another nodal set

We have $0 \leq \theta \leq \alpha$ and $0 \leq r \leq R$ (we can extend ϕ to be zero outside Ω and call the resulting function u). Now

$$-\int_0^R \frac{d}{dr}(u^2)\sinh r\,dr = \left. -u^2\sinh r\right|_0^R + \int_0^R u^2\cosh r\,dr .$$

Hence

$$\int_0^R u^2\sinh r\,dr \leq \int_0^R u^2\cosh r\,dr$$

$$= -\int_0^R \frac{d}{dr}(u^2)\sinh r\,dr$$

$$= -\int_0^R 2u\,u_r\sinh r\,dr$$

$$\leq 2\left(\int_0^R u^2\sinh r\,dr\right)^{1/2}\left(\int_0^R (u_r)^2\sinh r\,dr\right)^{1/2},$$

i.e.,

$$\int_0^R u^2 \sinh r\, dr \le 4 \int_0^R (u_r)^2 \sinh r\, dr.$$

Hence

$$\int_0^\alpha \int_0^R |\nabla u|^2 d\mu(r) \ge \int_0^\alpha \int_0^R (u_r)^2 \sinh r\, dr\, d\theta \ge \frac{1}{4} \int_0^\alpha \int_0^R u^2 \sinh r\, dr\, d\theta$$

which implies that

$$\lambda_1^{(N)} \ge 1/4.$$

In particular the above shows that

$$\inf_{\substack{\phi \text{ defined on } T \\ \int_T \phi d\mu = 0}} \frac{\int_T |\nabla \phi|^2 d\mu}{\int_T |\phi|^2 d\mu} \ge \frac{1}{4}.$$

This applies to $T = \Gamma(1)\backslash \mathbf{H}$ which, as in Figure 1.1, is a triangle. In particular it applies to f's which are $\Gamma(1)$ periodic, that is, we have

$$\lambda_1(\Gamma(1)\backslash \mathbf{H}) \ge 1/4$$

proving Proposition A.2.2. □

Remark A.2.4. The spectrum of Δ on $L^2(\Gamma\backslash\mathbf{H})$ is of fundamental interest. The eigenfunctions of Δ which are Γ periodic and square summable (other than the constant function) are known as Maaß forms and in this case are cusp forms. For these there is again the problem of estimating their Fourier coefficients and one has an analogue of the Ramanujan conjectures, though these are far from being solved at present. See [JL] for details.

To prove Theorem A.2.1 we introduce Poincaré series as in Section 1.5. This time they are adapted to $\Gamma\backslash\mathbf{H}$ and to Δ.

For $m > 0$ let

$$U_m(z, s) = \sum_{\gamma \in \Gamma_\infty\backslash\Gamma} y(\gamma z)^s e(m\gamma z). \qquad (A.2.4)$$

This series is clearly dominated by the corresponding Eisenstein series in (1.4.3) and hence converges absolutely in $\mathrm{Re}(s) > 1$. Moreover the term in the series corresponding to $\gamma = \pm 1$ is of the form $y^s e(mz)$ which is rapidly decreasing as $y \to \infty$. It follows from this and the properties of the Eisenstein series developed in Section 1.4 that $U_m(z, s)$ is in $L^2(\Gamma\backslash\mathbf{H})$ for $\mathrm{Re}(s) > 1$. In fact it is clear that

$$\|U_m(z, s)\|_2 = O(1) \qquad \text{uniformly for } \mathrm{Re}(s) \ge \sigma_0 > 1. \qquad (A.2.5)$$

Now

$$\Delta(y^s \, e(mz)) + s(1-s) \, (y^s \, e(mz)) = -4\pi m s (y^{s+1} \, e(mz)).$$

Hence for $\operatorname{Re}(s) > 1$ we have

$$(\Delta_z + s(1-s)) \, (U_m(z,s)) = -4\pi m s \, U_m(z, s+1). \qquad \text{(A.2.6)}$$

This may be written as

$$U_m(z,s) = -R_{s(1-s)}(\Delta) \, [4\pi m s \, U_m(z, s+1)], \qquad \text{(A.2.7)}$$

where R_λ is the resolvent $(\lambda - \Delta)^{-1}$ of Δ. We may use (A.2.7) to meromorphically continue the function $U_m(z,s)$ in s. In fact, restricting our attention to $\Gamma = \Gamma(1)$ we see from A.2.2 that $R_{s(1-s)}$ is analytic in $\operatorname{Re}(s) > 1/2$, with the only pole at $s = 1$ corresponding to $\lambda_0 = 0$. Hence from (A.2.7) (since the right hand side is shifted by 1) we see that

$$U_m(z,s) \qquad \text{is holomorphic in } \operatorname{Re}(s) > 1/2. \qquad \text{(A.2.8)}$$

The pole at $s = 1$ does not occur for $U_m(z,s)$ since it is easily seen from the definition of U that $\langle U_m(z,s), 1 \rangle \equiv 0$.

For a general selfadjoint operator Δ it is an easy consequence of the spectral theorem that

$$\|R_\lambda\| \le \frac{1}{\text{distance} \, (\lambda, \operatorname{spect} \Delta)}. \qquad \text{(A.2.9)}$$

Now $\text{dist} \, [s(1-s), \sigma(\Delta)] \ge |t| \, (2\sigma - 1)$, where $s = \sigma + it$ and $\sigma > 1/2$. Combining (A.2.9) with (A.2.7) and (A.2.5) we get

$$\|U_m(\,\cdot\,, s)\|_2 \ll \frac{|s| \, m}{|t| \, |2\sigma - 1|} \qquad \text{(A.2.10)}$$

for $\operatorname{Re}(s) \ge \sigma_0 > 1/2$.

The relation of $U_m(z,s)$ to Kloosterman sums comes, as usual, by computing the Fourier development of the $z \mapsto z + 1$ periodic function $U_m(z,s)$. In fact for $m, n > 0$ let

$$Z(m,n,s) = \sum_{c=1}^{\infty} \frac{K(m,n,c)}{c^{2s}} \qquad \text{(A.2.11)}$$

which we refer to as the Selberg–Kloosterman zeta function, see Selberg [Se]. By Weil's estimate it is clear that $Z(m,n,s)$ is holomorphic in $\operatorname{Re}(s) > 3/4$ and in fact

$$|Z(m,n,s)| \ll 1 \qquad \text{for } \sigma \ge \sigma_0 > 3/4. \qquad \text{(A.2.12)}$$

Lemma A.2.5.

$$\langle U_m(\,\cdot\,,s), \overline{U_n(\,\cdot\,,\overline{s}+2)}\rangle = 4^{-s-1}\,\pi^{-1}\,n^{-2}\,\frac{\Gamma(2s+1)}{\Gamma(s)\,\Gamma(s+2)}\,Z(m,n,s) + R(s)$$

where $R(s)$ is holomorphic in $\mathrm{Re}\,(s) > 1/2$ and satisfies

$$R(s) \ll O\left(\frac{1}{\sigma - 1/2}\right)$$

in this region.

Proof: By the usual calculation as in Section 1.5

$$\int_{\Gamma(1)\backslash \mathbf{H}} U_m(z,s)\,\overline{U_n(z,w)}\,\frac{dx\,dy}{y^2}$$

$$= \int_0^\infty \int_0^1 e(mz)\,y^s\,\overline{U_n(z,w)}\,\frac{dx\,dy}{y^2}$$

$$= \delta_{m,n}(4\pi n)^{1-s-\overline{w}}\,\Gamma(s+\overline{w}-1)$$

$$+ \sum_{c\neq 0}\frac{K(m,n,c)}{|c|^{2s}}\int_0^\infty\int_{-\infty}^\infty\frac{y^{\overline{w}-s}}{(x^2+1)^s}\,e\left[\frac{-m}{yc^2(x+1)} - n(xy-iy)\right]\frac{dx\,dy}{y}\,.$$

Now one may evaluate the integral

$$\int_{-\infty}^\infty (x^2+1)^{-s}\,e^{-2\pi m x}dx = \frac{-\pi(y\pi m)^{s-1}}{\Gamma(s)}\,W_{0,s-1/2}(4\pi m y)\,,$$

where $W_{0,\mu}(z)$ is the Whittaker function [GR, p. 860] and also

$$\int_0^\infty e^{-2\pi N y}y^s\,W_{0,\mu}(4\pi N y)\frac{dy}{y} = (4\pi N)^{-s}\,\frac{\Gamma(s+1/2+\mu)\,\Gamma(s+1/2-\mu)}{\Gamma(s+1)}\,.$$

Using these in the above and setting $w = \overline{s}+2$, we get

$$\langle U_m(\,\cdot\,,s), U_n(\,\cdot\,,\overline{s}+2)\rangle$$

$$= \delta_{m,n}(4\pi n)^{-2s-1}\Gamma(2s+1)$$

$$+ 4^{-s-1}\pi^{-1}n^{-2}\,\frac{\Gamma(2s+1)}{\Gamma(s)\Gamma(s+2)}\,Z(m,n,s) + \sum_{c\neq 0}\frac{K(m,n,c)}{|c|^{2s}}\,R_{m,n}(s,c)$$

$$\text{(A.2.13)}$$

where

$$R_{m,n}(s,c) = \int_0^\infty\int_{-\infty}^\infty\frac{y^2}{(x^2+1)^s}\left\{\exp\left(2\pi i m\,\frac{x-i}{yc^2(x^2+1)}\right) - 1\right\}$$

$$\times \exp(2\pi i n(xy-iy))\,\frac{dx\,dy}{y}\,.$$

Now

$$\int_0^\infty y \left| \exp\left\{ -2\pi m \frac{x-i}{yc^2(x^2+1)} \right\} - 1 \right| \exp(-2\pi n y)\, dy$$

$$\ll \int_0^{c^{-2}} y\, dy + \int_{c^{-2}}^\infty y\, \frac{\exp(-2\pi n y)}{c^2 y}\, dy$$

$$\ll c^{-2}.$$

Hence $R_{m,n}(s,c) \ll c^{-2}/|\sigma - 1/2|$.

This implies

$$\sum_{c \neq 0} \frac{K(m,n,c)}{|c|^{2s}} R_{m,n}(s,c)$$

is holomorphic in $\mathrm{Re}\,(s) > 1/2$ and is $O\left(\frac{1}{\sigma - 1/2}\right)$ in this region. This proves Lemma A.2.5.   □

As a consequence of (A.2.10) and Lemma A.2.5 we get the following estimate on the growth of the zeta function $Z(m,n,s)$

Theorem A.2.6. $Z(m,n,s)$ *is holomorphic in* $\mathrm{Re}\,(s) > 1/2$ *and*

$$|Z(m,n,s)| = O_{m,n}\left(\frac{|s|^{1/2}}{\sigma - 1/2} \right) \qquad \text{as } t \to \infty \text{ in } \sigma > 1/2.$$

Proof: This follows from (A.2.5), (A.2.10), Cauchy's inequality and Stirling's formula for the gamma function.   □

The passage from Theorem A.2.6 to A.2.1 is standard; we review it briefly. Using Theorem A.2.6 and A.2.1 and applying the Phragmén-Lindelöf principle [Ti] we have

$$\left| Z\left(m, n, \frac{1+s}{2} \right) \right| \ll |t|^{1/2 - \sigma + \epsilon}$$

for $\epsilon \leq \sigma \leq 1/2 + \epsilon$. Proceeding by the usual Mellin inversion [Da1] we get

$$\sum_{c \leq x} \frac{K(m,n,c)}{c} = \frac{1}{2\pi i} \int_{1/2 + \epsilon_0 - iT}^{1/2 + \epsilon_0 + iT} Z\left(m, n, \frac{s+1}{2} \right) \frac{x^s}{s}\, ds + O\left(\frac{x^{1/2 + \epsilon}}{T} \right).$$

Now shifting the contour to $\mathrm{Re}\,(s) = \epsilon$ and using Theorem A.2.6 we get

$$\sum_{c \leq x} \frac{K(m,n,c)}{c} = O\left(x^\epsilon T^{1/2 + \epsilon} + \frac{x^{1/2 + \epsilon}}{T} \right).$$

Letting $T = x^{1/3}$ gives Theorem A.2.1.   □

To conclude this Appendix we note that all of the above could be carried out in the same way for any congruence group Γ (not necessarily $\Gamma = \Gamma(1)$) except that we would not know (and do not know) that $\lambda_1(\Gamma\backslash\mathbf{H}) \geq 1/4$. In fact the above arguments yield in the general case by Weil's bound (1.5.7) that

$$\lambda_1(\Gamma\backslash\mathbf{H}) \geq \frac{3}{16} \qquad (A.2.14)$$

for any congruence subgroup Γ.

The bound (A.2.14) is due to Selberg [Se] and represents a step towards the 'Ramanujan' Conjecture A.2.3. It is the equivalent of Proposition 1.5.3. (A.2.14) has to date been improved only to the extent $\lambda_1(\Gamma\backslash\mathbf{H}) > 3/16$, see [GJ]. Iwaniec [IW3] shows that for almost all $(\Gamma_0(p), \chi)$ a further improvement is possible.

Notes and comments on Chapter 1

For the following comments we assume that the reader is familiar with the various notions from the theory of automorphic forms that are used without being defined.

As mentioned earlier the Ramanujan conjecture for forms of even integral weight k for $\Gamma(N)$ were solved by Deligne [De]. Previously, Eichler [Ei1] and Igusa [Ig] settled the case of $k = 2$. The solutions above reduce the problem to the Riemann Hypothesis for curves over finite fields when $k = 2$ and to their generalizations – the Weil conjectures – for $k \geq 4$. The Hecke operators play a central role here. For example when $k = 2$ the key comes from investigating their reduction $(\bmod\, p)$ and its relation to the Frobenius endomorphism. To obtain the Ramanujan conjectures in the form that we have stated in 1.2.5, one must also investigate the Hecke operators T_p for p dividing the level of Γ and in particular invoke the theory of new forms [AL]. See Rankin [Ra] for a treatment of what is needed here.

For the more general case of an arbitrary automorphic form, i.e., Maaß form [JL] as defined in A.2.4, as well as for automorphic forms on $GL(2)$ over a number field, the Ramanujan conjectures remain unsolved. The method of Section 1.5, which is due to Petersson [Pe], can be generalized and yields non-trivial, and in general, the best known estimates. A striking approach to the general Ramanujan conjectures in representation theory has been put forth by Langlands [L]. In this approach automorphic forms on other groups such as $GL(n)$ play a key role. In this context it should be pointed out that there is a very natural formulation of the Ramanujan conjectures in terms of representation theory due to Satake [Sa]. This formulation which asserts that for an automorphic cuspidal representation $\pi = \otimes_p \pi_p$, π_p for unramified p, is not in the complementary series. This viewpoint shows in particular that the Ramanujan conjectures 1.2.5 and the Selberg conjecture (A.2.3) are really one and the same. Moreover this formulation extends to other groups, though care must be exercised in doing so, as was discovered by Howe and Piatetski-Shapiro [HP]. They give examples similar to the 1/2-integral weight example 1.3.8 for which the natural Ramanujan conjecture fails. In general very little is proven in the direction of the Ramanujan conjectures. There are some qualitative results concerning the existence of a non-trivial estimate due to Kazhdan [Ka]. These are shared by groups which have 'property T' and in particular groups of rank ≥ 2 (we note that this 'property T' and the corresponding non-trivial estimate has nothing to do with Γ, the discrete subgroup, and hence apparently nothing to do with arithmetic). In cases of the remaining classical groups, which do not have property T, a quantitative result similar to Proposition 1.5.3 and (A.2.14), and using the same basic technique, has been established by Elströdt–Mennicke–Grünewald [EMG] and Li–Piatetski-Shapiro–Sarnak [LiPS] for $SO(n, 1)$ and by Li [Li] for $SU(n, 1)$.

Concerning the Ramanujan conjectures for half integral weight, a non-trivial estimate in that direction will be established in Chapter 4. This case appears to be fundamentally different from the even integral weight case and the conjecture is in fact related to some other deep conjectures. Waldspurger [Wa] and Konen–Zagier [KZ] have shown that for say $\Gamma_0(4)$, if $f(z) \in S_k(\Gamma_0(4))$ is an eigenform for the Hecke operators T_{p^2} and satisfies a further orthogonality condition, and if g is the Shimura correspondent to f, so that g is an eigenform in $S_{2k-1}(\Gamma(1))$, then

$$\frac{|a_{|D|}|^2}{\langle f, f \rangle} = \frac{(k - 3/2)!}{\pi} \frac{|D|^{k-1}}{\langle g, g \rangle} L(g \otimes \chi_D, k - 1/2).$$

Here $D > 0$ is a fundamental discriminant, the a_n's are the Fourier coefficients of f and $L(g \otimes \chi_D, s)$ is the automorphic L function attached to $g \otimes \chi_D$. The special value $s = k - 1/2$ is the middle of the critical strip of $L(g \otimes \chi_D, s)$. What emerges is that the Ramanujan conjecture for f is equivalent to the Lindelöf hypothesis for L (in the D aspect at the point $k - 1/2$)! Needless to say, it would be of great interest to develop, even conjecturally, as in the Langlands program mentioned above, an approach to the Lindelöf hypothesis. For some interesting work in this direction see Bump–Hoffstein–Friedberg [BHF].

The theta function construction of automorphic forms in Section 1.3 was generalized by Siegel [Si1] to include arbitrary quadratic forms. The general setting and score for this theory has been laid down by Weil [We2] and Howe [H1] respectively; see the latter's theory of dual pairs.

The main Theorem A.2.1 of Appendix 2 is due to Kuznietzov [Ku]. The proof of this result by use of Selberg's zeta function and especially the growth estimate Theorem A.2.6 follows the paper of Goldfeld and Sarnak [GS]. The foundational aspects of Appendix 1.2 (and also an excellent discussion of the problem of estimating Fourier coefficients) is due to Selberg [Se]. As was remarked in the text, this theory in the more flexible form of what is known as the Kuznietsov Trace Formula [Ku] has been spectacularly applied by Deshouillers and Iwaniec [DI]. The analog of Theorem A.2.1 for forms of more general weight has been used by Vardi [V] to prove the equidistribution of Dedekind sums. Meyerson [Mey], using Vardi's results, was able to settle a related conjecture of Rademacher concerning the equidistribution of such sums.

Chapter 2

Invariant Means on $L^\infty(S^n)$

2.1 Invariant means

Let $X = L^\infty(S^n)$ denote the Banach space of essentially bounded measurable functions on the n–sphere with respect to the Lebesgue measure λ. For $f \in X$ and $t \in SO(n+1)$, a rotation, let $f_t(x) = f(tx)$. By duality we may define $\nu \in X^*$ by $\nu_t f = \nu(f_t)$. An invariant mean on X, call it ν, is a linear functional satisfying

$$\nu(1) = 1 \qquad\qquad\qquad\qquad (i)$$

$$\nu(f) \geq 0 \qquad \text{if } f \geq 0 \qquad\qquad\qquad (ii)$$

$$\nu_t = \nu \qquad \text{for all } t \in SO(n+1). \qquad\qquad (iii)$$

λ is an invariant mean and, as is well known, the only one when viewed as an element of $C(S^n)^*$ (i.e., as a measure). The problem we address here is that of the uniqueness of λ as an invariant mean on X. As was pointed out in the Introduction, for $n \geq 2$, this is the Ruziewicz problem [Ba]. In Section 2.2 we show that for $n = 1$, λ is far from being unique. In Section 2.3 the question of uniqueness is shown to follow from the existence of an 'ε–good set' whose definition is as follows:

A set $t_1, \ldots, t_r \in SO(n+1)$ is said to be ε–good, where $\varepsilon > 0$ is a fixed number, if for any $f \in L^2(S^n)$ with $\int_{S^n} f d\lambda = 0$ there is a $j \in \{1, \ldots, r\}$ such that

$$\|f_{t_j} - f\|_2 \geq \varepsilon \|f\|_2 .$$

Thus the problem of uniqueness for $n \geq 2$ becomes one of constructing such an ε–good set. In Section 2.4 we show how to inductively (and effectively) construct an ε'–good set in $SO(n+2)$ from an ε–good set in $SO(n+1)$.

Finally in Section 2.5 an explicit and even optimal ε–good set is constructed in $SO(3)$. It is in the proof of the 'ε–goodness' that we make use

45

of the Ramanujan conjectures.

An ε–good set clearly generates a superbly ergodic group $\Gamma \subset SO(n + 1)$. As such the explicit ε–good sets can be used to generate optimally equidistributed matrices in orthogonal groups and hence such points on a sphere. We explain these ideas briefly in Section 2.6. For the record our basic ε–good set is the set $t_1, t_2, t_3 \in SO(3)$ of rotations about the x_1, x_2, x_3 axes in $\mathbf{R}^3$ through an angle of $\arccos(-3/5)$ – this set is $\frac{2}{3}(3 - \sqrt{5})$–good.

2.2   Nonuniqueness for $L^\infty(S^1)$

Let H be the linear subspace of $X = L^\infty(S^1)$ spanned by functions of the form $f_t - f$ where $t \in S^1$ and $f \in X$. Clearly $\nu \in X^*$ is invariant iff it annihilates H. To produce an invariant mean other than λ we need to find elements in X outside of $\overline{H}$ other than constants.

Lemma 2.2.1. Let $A \subset S^1$ be a fixed open dense set, then for any $h \in H$

$$\operatorname{ess\,inf}_{x \in A} h(x) \leq 0.$$

Proof: Let $h \in H$. h is of the form $\sum_{k=1}^{N}(f_k)_{t_k} - f_k$. For a large number M consider

$$T(x) = \sum_{|m| \leq M} h(x + m_1 t_1 + \ldots + m_N t_N)$$

$$= \sum_{k=1}^{N} \sum_{|m| \leq M} [(f_k)_{t_k}(x + m_1 t_1 + \ldots + m_N t_N)$$
$$- f_k(x + m_1 t_1 + \ldots + m_N t_N)]$$

where $m \in \mathbf{Z}^N$. Clearly for any x

$$|T(x)| \leq N\,M^{N-1} \sum_{k=1}^{N} \|f_k\|_\infty. \qquad (2.2.1)$$

On the other hand if $\operatorname{ess\,inf}_{x \in A} h(x) = \varepsilon$, then since A is open and dense we can clearly choose x so that the points $x_1 + m_1 t_1 + \ldots + m_N t_N$, $|m| \leq M$ are all in A and hence for such x

$$T(x) \geq M^N \varepsilon. \qquad (2.2.2)$$

Since M is arbitrarily large, it follows from (2.2.1) and (2.2.2) that $\varepsilon \leq 0.\square$

With this Lemma it is easy to see that $H + \mathbf{R}\chi_{S^1} + \mathbf{R}\chi_{A^c} \triangleq Y$ is a direct sum. In fact, define the linear functional ν on Y by

$$\nu(h + \alpha\chi_{S^1} + \beta\chi_{A^c}) = \alpha \qquad (2.2.3)$$

where we now assume that A has been chosen so that $\lambda(A^c) \neq 0$. Then

$$\|h + \alpha\chi_{S^1} + \beta\chi_{A^c}\| \geq \operatorname{ess\,sup}_{x \in A} |h + \alpha\chi_{S^1}| = |\alpha|$$

by Lemma 2.2.1, i.e.,

$$|\nu(y)| \leq \|y\| \qquad \text{for } y \in Y. \qquad (2.2.4)$$

By the Hahn–Banach theorem we may extend ν to a linear functional ν on X satisfying

(i) $\qquad\qquad\qquad\qquad\qquad \nu(H) = 0$

(ii) $\qquad\qquad\qquad\qquad\qquad \nu(1) = 1$

(iii) $\qquad\qquad\qquad\qquad\qquad \|\nu\| = 1$

(iv) $\qquad\qquad\qquad\qquad\qquad \nu(\chi_{A^c}) = 0$

These imply that ν is an invariant mean and (iv) asserts that $\lambda \neq \nu$.

The reader familiar with the notion of an amenable group will recognize that it is this property that is crucial in deriving (2.2.1). Not suprisingly the uniqueness of λ for non–discrete, amenable (as discrete), G always fails [Gr,Ru]. $G = SO(3)$ is not amenable as a discrete group. In fact, as we will see later, the rotations t_1, t_2, t_3 mentioned at the end of the Introduction to Chapter 2 form a free group.

2.3   Reduction to ε–good sets

Theorem 2.3.1. *If $t_1, \ldots, t_k \in SO(n + 1)$ are an ε–good set then the Lebesgue measure is the unique invariant mean on $L^\infty(S^n)$.*

Proof: Let $\nu \in X^*$ be an invariant mean. Now $X = (L^1)^*$ and hence $\nu \in (L^1)^{**}$. L^1 is weak–star dense in $(L^1)^{**}$ and since $\nu \geq 0$ and $\nu(\chi_{S^n}) = 1$ we can find a net $\{f^{(\mu)}\} \in L^1$ with $f^{(\mu)} \to \nu$ weak–star, $\int f^{(\mu)}d\lambda = 1$ and $f^{(\mu)} \geq 0$. Moreover since ν is invariant we have that for each $j = 1, \ldots, k$, $f_{t_j}^{(\mu)} - f^{(\mu)} \to 0$ weakly in L^1 (since it does so weak–star in $(L^1)^{**}$). The weak and strong closures of convex sets coincide so that we may take convex combinations of the tails of $f^{(\mu)}$ to get a new net $g^{(\mu)}$ for which $g_{t_j}^{(\mu)} - g^{(\mu)} \to 0$

strongly in L^1 for each j, and also $\int g^{(\mu)} d\lambda = 1$ and $g^{(\mu)} \geq 0$. Moreover, $g^{(\mu)} \to \nu$ weak–star. It follows that if $\sqrt{g^{(\mu)}} = h^{(\mu)}$ then

$$\|h_{t_j}^{(\mu)} - h^{(\mu)}\|_2^2 \leq \|g_{t_j}^{(\mu)} - g^{(\mu)}\| \to 0$$

for each j. Thus we have $\|h^{(\mu)}\|_2 = 1$ and $\|h_{t_j}^{(\mu)} - h^{(\mu)}\|_2 \to 0$. The ε–good property ensures then that $h^{(\mu)} \to 1$ in L^2 and hence that

$$\|g^{(\mu)} - 1\|_1 \leq \|h^{(\mu)} - 1\|_2 \, \|h^{(\mu)} + 1\|_2 \to 0 \,.$$

That is $g^{(\mu)} \to 1$ strongly in L^1 and since also $g^{(\mu)} \to \nu$ weak–star $\Rightarrow \nu = 1$, completing the proof.　　　　　　　□

It is clear that a set $R_1, \ldots, R_k$ in $SO(n+1)$ is at most 2–good. Actually it cannot be as good as $\sqrt{2}$–good.

Proposition 2.3.2. *A finite set $R_1, \ldots, R_k \in SO(n)$, $n \geq 3$, is at most $\sqrt{2}$ good.*

Proof: (C. McMullen) Let $R_1, \ldots, R_k$ be a given set of rotations. We can clearly find a small open set U such that $(R_j U) \cap U = \emptyset$ for each j. Now let f be supported in U with $\int_{S^{n-1}} f \, d\lambda = 0$ and $\int_{S^{n-1}} |f^2| d\lambda = 1$. Then

$$\langle R_j f, f \rangle = 0 \qquad \text{for } j = 1, \ldots, k \,,$$

and so

$$\|R_j f - f\|_2^2 = 2 \qquad \text{for } j = 1, \ldots, k \,.　　□$$

In Section 2.5 we will construct $(\sqrt{2} - \eta)$–good sets for every $\eta > 0$.

2.4　Inductive construction

Suppose we are given an ε–good set $t_1, \ldots, t_k \in SO(n+1)$. We construct $2k$ rotations in $SO(n+2)$ as follows:
For each $j = 1, \ldots, k$ let

$$\tilde{t}_j = \begin{pmatrix} t_j & \begin{matrix} 0 \\ 0 \end{matrix} \\ 0 \ 0 & 1 \end{pmatrix} \in SO(n+2) \,. \qquad (2.4.1)$$

Thus $\tilde{t}_j$ fixes the x_{n+2} axis in $\mathbf{R}^{n+2}$, that is it is a rotation about the N–S axis of S^{n+1}. In a similar way we define $\tilde{s}_j \in SO(n+2)$, $j = 1, \ldots, k$ by making the same construction but with the N–S axis replaced by the 'E–W axis' (by which we mean any axis orthogonal to the N–S axis).

Theorem 2.4.1. *$\tilde{t}_1, \ldots, \tilde{t}_k, \tilde{s}_1, \ldots, \tilde{s}_k$ are $\varepsilon/(2k)$–good.*

Proof: As coordinates for S^{n+1} we use spherical coordinates (θ, τ), $0 \leq \theta < \pi$, $\tau \in S^n$. θ is measured from the x_{n+2} axis. In these coordinates

$$f_{\tilde{t}_j}(\theta, \tau) = f(\theta, t_j \tau).\tag{2.4.2}$$

A similar statement holds for the $\tilde{s}_j$ if spherical coordinates about the E–W axis are used. Define subspaces of $L^2(S^{n+1})$, A_N, B_N, A_E, B_E by

$$A_N = \{f\mid f(\theta, \tau) = f(\theta, P\tau),\ \forall P \in SO(n+1)\},\tag{2.4.3}$$

i.e., A_N consists of the functions radial about the N–S axis. $B_N = A_N^\perp$ while A_E and B_E are defined similarly.

For $f \in B_N$ and $g \in B_E$ we claim that

$$\sum_{j=1}^{k} \|f_{\tilde{t}_j} - f\|_2 \geq \varepsilon \|f\|_2$$

$$\sum_{j=1}^{k} \|g_{\tilde{s}_j} - g\|_2 \geq \varepsilon \|g\|_2\tag{2.4.4}$$

In fact $f \in B_N \Rightarrow \int_{S^n} f(\theta, \tau)\, d\lambda_n(\tau) = 0$ for almost all θ. For each such θ (2.4.2) and the hypothesis that $t_1, \ldots, t_k$ is ε-good in τ, we find that an integration in θ leads to (2.4.4). The same goes for B_N and $\tilde{s}_j$.

Before continuing with the proof we establish the following important lemma.

Lemma 2.4.2. *If* $f \in A_N$, $g \in A_E$ *and* $\int_{S^{n+1}} f d\lambda = 0 = \int_{S^{n+1}} g\, d\lambda$ *then*

$$|\langle f, g\rangle| \leq \frac{1}{n+1}\, \|f\|_2 \|g\|_2\,.$$

Proof: Decompose $L^2(S^{n+1})$ into the orthogonal decomposition in spherical harmonics

$$L^2(S^{n+1}) = \bigoplus_{k=0}^{\infty} H_k$$

where H_k is the space of spherical harmonics of degree k. We need only check the Lemma for $f, g \in H_k$, $k \geq 1$ and $\|f\|_2 = \|g\|_2 = 1$. Moreover such an $f \in H_k \cap A_N$ (and $g \in H_k \cap A_E$) necessarily lies in the 1–dimensional space of N–S rotation invariant eigenfunctions of the Laplacian on S^{n+1}. That is $f(z) = \mu W_k(z, N)$, where $W_k(z, \zeta)$ is the zonal function [Fo] normalized so that $W_k(z, z) = 1$, and μ is a scalar chosen so that $\langle f, f\rangle = 1$. Similarly

$g(z) = \mu\, W_k(z, E)$. Thus

$$\langle f, g \rangle = \int_{S^n} f(z)\,\overline{g(z)}d\lambda = \mu^2 \int_{S^{n+1}} W_k(z, N)\, W_k(z, E)\, d\lambda(z)$$

$$= \mu^2\, h(k)\, W_k(N, E) \tag{2.4.5}$$

where h depends on k only; the 'Selberg transform' $W(z, \zeta)$ is a spherical function (see [Sz] or simply compute).

Moreover we also have

$$1 = \langle f, f \rangle = \mu^2\, h(k)\, W_k(N, N) = \mu^2\, h(k)\,. \tag{2.4.6}$$

Thus $\langle f, g \rangle = W_k(N, E)$, and the Lemma will follow if we can show that

$$\sup_{k \geq 1} |W_k(N, E)| = \frac{1}{n+1}\,. \tag{2.4.7}$$

It is well known that the zonal function $W_k(z, N)$ as a function of $x = \cos\theta$ is a multiple of the ultra spherical polynomial $P_k^{(n-2)/2}$, [Fo] and [Sz]. Since the N–S and E–W axes are orthogonal, what we want is the value of P at $x = 0$. This may be evaluated explicitly [Sz] giving

$$W_k(N, E) = \frac{\Gamma\left(2k + \dfrac{n+1}{2}\right)\, \Gamma(k)\, \dbinom{k - 1/2}{k}}{\Gamma\left(k + \dfrac{n+1}{2}\right)\, \Gamma(2k+1)\, \dbinom{2k + (n-1)/2}{2k}}\,.$$

The maximum occurs for $k = 1$ giving the Lemma.   □

We can now complete the proof of Theorem 2.4.1.

Given f with $\int_{S^{n+1}} f\, d\lambda = 0$ and $\|f\|_2 = 1$ we may write

$$f = a_N + b_N \qquad \text{and also} \quad f = a_E + b_E \tag{2.4.8}$$

with $a_N \in A_N$, $b_N \in B_N$, $a_E \in A_E$, $b_E \in B_E$. The considerations above show that

$$\|f_{\tilde{t}_j} - f\|_2 \geq \frac{\varepsilon}{k}\,\|b_N\|_2 \qquad \text{for some } j$$

$$\|f_{\tilde{s}_{j'}} - f\|_2 \geq \frac{\varepsilon}{k}\,\|b_N\|_2 \qquad \text{for some } j'$$

We also have

$$\|a_N - a_E\|_2^2 = \|b_N - b_E\|_2^2$$

and

$$\|a_N - a_E\|_2^2 \geq \frac{2n}{n+1}\,\|a_N\|_2\,\|a_E\|_2$$

by Lemma 2.4.2. Hence

$$\frac{2n}{n+1} \|a_N\|_2 \|a_E\|_2 \le (\|b_N\|_2 + \|b_E\|_2)^2 .$$

The last combined with $\|a_N\|_2^2 + \|b_N\|_2^2 = \|a_E\|_2^2 + \|b_E\|_2^2 = 1$ leads to the fact that one of $\|b_N\|_2$ or $\|b_E\|_2$ is $\ge 1/2$. From (2.4.9) this means that $\tilde{t}_1, \ldots, \tilde{t}_k, \tilde{s}_1, \ldots, \tilde{s}_k$ are $\varepsilon/(2k)$–good. $\qquad \square$

2.5 ε–good sets for $SO(3)$

In this section we give explicit and optimal ε–good sets in $SO(3)$. We know from Sections 2.2 and 2.3 that there can be no ε–good sets in S^1. It is instructive to see this directly. Let $\alpha_1, \ldots, \alpha_r \in S^1$. We may choose $0 \ne m \in$ **Z** such that $m(\alpha_1, \ldots, \alpha_r)$ (mod 1) in $\mathbf{R}^r / \mathbf{Z}^r$ is close to $(0, \ldots, 0) \in \mathbf{R}^r / \mathbf{Z}^r$. This follows the pigeon hole principle. It follows that for $\varepsilon > 0$ there is an m such that the function

$$f(x) = e(mx) \qquad \text{on } S^1$$

satisfies $\int_{S^1} f \, d\lambda = 0$ and $\|f_{\alpha_j} - f\|_2 = \|1 - e(m\alpha_j)\| < \varepsilon$ for each j. Thus it is clear that ε–good sets cannot exist in S^1.

For $SO(3)$ they exist and together with the inductive scheme of Section 2.4 these can then be used to construct explicit ε–good sets for $SO(n)$, $n \ge 4$ as well. Combined with Theorem 2.3.1 this establishes the uniqueness of λ for S^n, $n \ge 2$. Precisely, we show that the rotations t_1, t_2, t_3 introduced at the end of Section 2.1 are $\frac{2}{3}(3 - \sqrt{5})$–good.

If $S \subset SO(3)$ is a finite symmetric set, i.e., $s \in S \Rightarrow s^{-1} \in S$ then we may define the 'Hecke operator' T_S on $L^2(S^2)$ by

$$T_S f(x) = \sum_{s \in S} f(s\, x) . \tag{2.5.1}$$

This is a Hecke operator of the type introduced in Section 1.6 (with $\Gamma =$ identity). T_S is a symmetric operator on $L^2(S^2)$ with spectrum clearly contained in $[-k, k]$ where $k = |S|$. k is in fact in spectrum (T_S) since it is the eigenvalue of T corresponding to the constant function. Let $\lambda_1(T_S)$ denote the absolute value of the next-to-largest eigenvalue of T_S. The similarity to Appendix A.2 is no coincidence.

Let

$$S_5 = \{t_1, t_1^{-1}, t_2, t_2^{-1}, t_3, t_3^{-1}\}$$

where t_1, t_2, t_3 are rotations about the x_1, x_2, x_3 axes through $\arccos(-3/5)$.

Theorem 2.5.1.

$$\lambda_1(T_s) \le 2\sqrt{5} .$$

Remark: In fact one can show, see Lubotzky–Phillips–Sarnak [LPS1] that

$$\text{spectrum}\,(T_S) = \{6\} \cup [-2\sqrt{5}, 2\sqrt{5}]\,.$$

From the variational characterization of the eigenvalues of a symmetric matrix it follows from Theorem 2.5.1 that

$$\|f_{t_1} - f\|_2^2 + \|f_{t_2} - f\|_2^2 + \|f_{t_3} - f\|_2^2 \geq 2\,(3 - \sqrt{5})\,\|f\|_2^2$$

for all $f \in L^2(S^2)$ with $\int f\,d\lambda = 0$. This implies that t_1, t_2, t_3 are $\frac{2}{3}\,(3-\sqrt{5})$–good. □

The set S_5 and the operator T_{S_5} comes naturally from integral quaternions. Let $H = \{\alpha = a_0 + a_1\mathbf{i} + a_2\mathbf{j} + a_3\mathbf{k}\}$, $\mathbf{i}^2 = \mathbf{j}^2 = \mathbf{k}^2 = -1$, $\mathbf{i}\mathbf{j} = -\mathbf{j}\mathbf{i} = \mathbf{k}$, etc., be the ring of quaternions. Then $H(\mathbf{Z})$ denotes the ring of integral quaternions, i.e., the ones with $a_j \in \mathbf{Z}$. For $\alpha \in H(\mathbf{Z})$, $\overline{\alpha} = a_0 - a_1\mathbf{i} - a_2\mathbf{j} - a_3\mathbf{k}$, is its conjugate and $N(\alpha) = \alpha\overline{\alpha}$. For $\alpha \in H(\mathbf{Z})$, $N(\alpha) \in \mathbf{Z}$. It is clear that the units of $H(\mathbf{Z})$ are precisely those $\alpha \in H(\mathbf{Z})$ with $N(\alpha) = 1$ and consist of the quaternions $\pm 1, \pm\mathbf{i}, \pm\mathbf{j}, \pm\mathbf{k}$. The number of $\alpha \in H(\mathbf{Z})$ with $N(\alpha) = n$ is equal to the number $r_4(n)$, using the notation of (1.1.4). We saw in (1.1.6) that

$$r_4(n) = 8 \sum_{\substack{d|n \\ 4\nmid d}} d\,.$$

Let p be a prime, $p \equiv 1 \pmod 4$. We consider the set of all $\alpha \in H(\mathbf{Z})$ with $N(\alpha) = p$. It is clear that for such an $\alpha = a_0 + a_1\mathbf{i} + a_2\mathbf{j} + a_3\mathbf{k}$, precisely one of the a_j's is odd. The units act on this set and it is easy to see that each α' has a unique associate $\alpha = \varepsilon\,\alpha'$ for which

$$N(\alpha) = p, \;\; \alpha \equiv 1 \pmod 2 \;(\text{in } H(\mathbf{Z})) \text{ and } a_0 > 0. \tag{2.5.2}$$

In view of (1.1.6) and the fact that there are eight units we see that the set of α satisfying (2.5.2) consists precisely of $p+1$ elements and it clearly splits into $\sigma = (p+1)/2$ conjugate pairs. Thus the set S_p of α satisfying (2.5.2) is of the form

$$S_p = \{\alpha_1, \overline{\alpha_1}, \ldots, \alpha_\sigma, \overline{\alpha_\sigma}\}\,. \tag{2.5.3}$$

There is a homomorphism of $H(\mathbf{R})^*$ (the invertible elements of $H(\mathbf{R})$) into $SU(2)$ given by

$$\alpha = a_0 + a_1\mathbf{i} + a_2\mathbf{j} + a_3\mathbf{k} \longrightarrow \frac{1}{\sqrt{N(\alpha)}}\begin{pmatrix} a_0 + a_1 i & a_2 + a_3 i \\ -a_2 + a_3 i & a_0 - a_1 i \end{pmatrix}. \tag{2.5.3}$$

The elements of $SU(2)$ correspond via stereographic projection to rotations in $SO(3)$. The homomorphism (2.5.3) allows us to think of $\alpha \in H(\mathbf{R})^*$ as an element of $SU(2)$ or $SO(3)$, which we do (the meaning will be clear from the context).

We define the Hecke operators T_n, $n \geq 1$ an integer; $T_n : L^2(S^2) \to L^2(S^2)$ by

$$T_n f(\zeta) = \frac{1}{2} \sum_{\substack{\alpha \equiv 1 \,(\mathrm{mod}\, 2) \\ N(\alpha) = n}} f(\alpha \zeta). \qquad (2.5.6)$$

A simple calculation shows that

$$S_5 = \{1 + 2\mathbf{i}, 1 - 2\mathbf{i}, 1 + 2\mathbf{j}, 1 - 2\mathbf{j}, 1 + 2\mathbf{k}, 1 - 2\mathbf{k}\}.$$

Via (2.5.3) this set S_5 and the earlier one (before Theorem 2.5.1) coincide and also T_5 as defined in (2.5.6) is exactly the T_{S_5} from before. Theorem 2.5.1 is thus a special case of

Theorem 2.5.2.

$$\lambda_1(T_p) \leq 2\sqrt{p}.$$

We will need a few Lemmata.

Lemma 2.5.3. *Every* $\beta \in H(\mathbf{Z})$ *with* $N(\beta) = p^k$ *has a unique representation*

$$\beta = p^\ell \varepsilon \, R_m(\alpha_1, \ldots, \overline{\alpha_\sigma})$$

where $\ell \leq k/2$, $m + 2\ell = k$, R_m *is a reduced word of length* m *in* $\alpha_1, \ldots, \overline{\alpha_\sigma}$ *and* ε *is a unit. By a reduced word in* $\alpha_1, \overline{\alpha_1}, \ldots, \overline{\alpha_\sigma}$ *we mean a word in these letters in which no* $\alpha_j \overline{\alpha_j}$ *or* $\overline{\alpha_j} \alpha_j$ *appears.*

Proof: We begin with the proof of the existence of such a factorization. In Dickson [Di] it is shown that $H(\mathbf{Z})$ is a left and right Euclidean ring and that an odd element of $H(\mathbf{Z})$ (i.e., $N(\alpha)$ is odd) is prime iff its norm is prime. Since $N(\beta) = p^k$, β is odd. We may therefore write $\beta = \gamma \delta$ where $N(\gamma) = p^{k-1}$ (if β were not already prime) and $N(\delta) = p$. By using a unit ε and the definition of the set S_p we have the expression

$$\beta = \gamma \varepsilon \alpha \qquad \text{with } \alpha \in S_p.$$

Repeating this factorization by factoring γ and carrying out the cancellation along the way leads to the required representation of β in Lemma 2.5.3.

To prove the uniqueness we count the number of such representations. It is easily seen that the number of reduced words of length $\ell \geq 1$ is

$$(p+1)\, p^{\ell-1}. \qquad (2.5.7)$$

Hence the number of factorizations of elements of norm p^k is

$$8 \left(\sum_{0 \leq \ell < k/2} (p+1)\, p^{k-2\ell-1} + \delta(k) \right) \qquad (2.5.8)$$

where $\delta(k) = 0$ if k is odd and $\delta(k) = 1$ if k is even. Summing the series we see that there are $8\,(p^{k+1} - 1)/(p - 1)$ such factorizations. On the other hand from (1.1.6) this is precisely the number of elements of norm p^k. Thus each representation is unique. $\square$

Lemma 2.5.4. *If* $\beta \equiv 1 \pmod 2$ *with* $N(\beta) = p^k$ *then* β *is uniquely expressible in the form*

$$\beta = \pm p^\ell\, R_m(\alpha_1, \ldots, \overline{\alpha_\sigma})$$

where $2\ell + m = k$ *and* R_m *is reduced.*

Proof: If $\beta \equiv 1 \pmod 2$ then since $\alpha_i \equiv 1 \pmod 2$ we see that in Lemma 2.5.3 $\varepsilon \equiv 1 \pmod 2$ whence $\varepsilon = \pm 1$. $\square$

We note that this Lemma ensures that $\alpha_1, \ldots, \alpha_\sigma$ when viewed as elements of $SO(3)$ generate a free group (see the end of Section 2.2 where this was quoted).

Lemma 2.5.5.

$$T_{p^\nu} = p^{\nu/2}\, U_\nu\!\left(\frac{T_p}{2\sqrt{p}}\right)$$

where U_ν *is the Chebyshev polynomial of the second kind*

$$U_\nu(x) = \frac{\sin((\nu + 1)\arccos x)}{\sin(\arccos x)}.$$

The proof is a straightforward application of Lemma (2.5.4). From that one derives the relations as in (3.4.33) of the next Section (as well as (3.4.35)). Note in particular if u is an eigenfunction of T_p say,

$$T_p u = \lambda u \qquad\qquad \text{then}$$

$$T_{p^\nu} u = p^{\nu/2}\, \frac{\sin((\nu + 1)\theta)}{\sin \theta} \qquad\qquad \text{where} \qquad (2.5.9)$$

$$\cos \theta = \frac{\lambda}{2\sqrt{p}}$$

To study the spectrum of T_n we examine its action on spherical harmonics of degree m on S^2, i.e., its action on $H_m(S^2)$. Clearly these finite dimensional subspaces are invariant for any Hecke operator T_S. Let $u \in H_m(S^2)$, $m \geq 1$.

Lemma 2.5.6. *Fix* $\zeta_0 \in S^2$ *then the function* $F(z)$ *for* $z \in \mathbf{H}$ *given by*

$$F(z) = \sum_{\substack{\alpha \equiv 2 \,(\mathrm{mod}\,4) \\ \alpha \in H(\mathbf{Z})}} N(\alpha)^m\, u(\alpha\,\zeta_0)\, e^{2\pi i N(\alpha) z / 32}$$

is a holomorphic cusp form of weight $2+2m$ for the congruence group $\Gamma(16)$.

Proof: We have seen in Section 1.3.3 (and in particular 1.3.7) that with

$$\dot{A} = \begin{pmatrix} 1 & & & \\ & 1 & & \\ & & 1 & \\ & & & 1 \end{pmatrix}, \; n = 4, \; N = 4 \text{ and } h = (2,0,0,0)$$

$$\tilde{\theta}(z, h, N) = \sum_{m \equiv h \;(\mathrm{mod}\, N)} P(m) \, e^{2\pi i |m|^2 z/16}$$

is a modular cusp form of weight $2 + \nu$ for $\Gamma(16)$, this being so for any spherical harmonic P of degree $\nu \geq 1$ in four variables. Thus to prove the Lemma it suffices to show that $u(\alpha \, \zeta_0) \, N(\alpha)^m$ with $\alpha = a + b\mathbf{i} + c\mathbf{j} + d\mathbf{k}$ is such a spherical harmonic in (a, b, c, d).

Without loss of generality we may assume that ζ_0 is the south pole in S^2. A simple calculation then gives

$$u(\alpha\zeta_0) \, N(\alpha)^m$$
$$= N(\alpha)^m \, u\left(\frac{1}{N(\alpha)} \, (2(ca - db), 2(da + bc), c^2 + d^2 - a^2 - b^2) \right).$$

Now $u(\zeta)$ is the restriction to the unit sphere of a homogeneous harmonic polynomial of degree m in three variables. Such a polynomial can be written as a sum of polynomials of the form

$$(\xi_1 x_1 + \xi_2 x_2 + \xi_3 x_3)^m \qquad \text{with } \sum_{j=1}^{3} \xi_j^2 = 0,$$

see [Sc2]. We may therefore assume that u is of this form. Then

$$N(\alpha)^m u(\alpha \, \zeta_0) = (2\xi_1(ca - db) + 2\xi_2(ad + bc) + \xi_3(c^2 + d^2 - a^2 - b^2))^m.$$

This is clearly a homogeneous polynomial of degree $2m$ in a, b, c, d. It is also harmonic as is easily checked using $\sum_{j=1}^{3} \xi_j^2 = 0$. This then proves Lemma 2.5.6. □

We now complete the proof of Theorem 2.5.2. $F(z)$ may be written as

$$F(z) = \sum_{\nu=1}^{\infty} a_\nu e\left(\frac{\nu z}{32}\right)$$

where

$$a_\nu = \nu^m \sum_{\substack{\alpha \equiv 2 \;(\mathrm{mod}\, 4) \\ N(\alpha) = \nu}} u(\alpha\zeta).$$

We apply the Ramanujan bound for cusp forms of weight $2 + 2m$ to get

$$|a_\nu| \ll_\varepsilon \nu^{m+1/2+\varepsilon}. \qquad (2.5.10)$$

(Here u and ζ are fixed and the implied constant depends on them.) Hence

$$\left| \sum_{\substack{\alpha \equiv 2 \,(\mathrm{mod}\,4) \\ \alpha \in H(\mathbf{Z}) \\ N(\alpha)=\nu}} u(\alpha\zeta) \right| \ll_\varepsilon \nu^{1/2+\varepsilon}. \qquad (2.5.11)$$

Writing $\mu = \nu/4$ (ν divisible by 4) and $\beta = \alpha/2$ in the above sum, gives $\beta \in H(\mathbf{Z})$, $N(\beta) = \mu$ and $\beta \equiv 1 \,(\mathrm{mod}\,2)$. Hence (2.5.11) reads

$$\left| \sum_{\substack{\beta \equiv 1 \,(\mathrm{mod}\,2) \\ N(\beta)=\mu}} u(\beta\zeta) \right| \ll_\varepsilon \mu^{1/2+\varepsilon}.$$

In particular for $\mu = p^k$, $p \equiv 1 \,(\mathrm{mod}\,4)$ we have using Lemma 2.5.5

$$|T_{p^k}u(\zeta)| = p^{k/2} \left| U_k\left(\frac{T_p}{2\sqrt{p}}\right) u(\zeta) \right| \ll_\varepsilon p^{k/2+\varepsilon k}.$$

Now if $T_p u = \lambda u$ and ζ is chosen so that $u(\zeta) \neq 0$ then the above combined with 2.5.9 gives

$$\left| \frac{\sin((k+1)\,\theta)}{\sin\theta} \right| \ll_\varepsilon p^{\varepsilon k}$$

where $\lambda = 2\sqrt{p}\cos\theta$. The last clearly implies θ is real and hence that $|\lambda| \le 2\sqrt{p}$. This proves Theorem 2.5.2.   □

Note that by using the bound established in Proposition 1.5.3 we would get

$$\lambda_1 \le p^{3/4} + p^{-3/4}$$

which, even though it does not capture the complete truth, does suffice for the application to the construction of an ε-good set. That is, it shows that t_1, t_2, t_3 are $\varepsilon = [6 - (5^{3/4} + 5^{-3/4})]$-good.

Recall that in Proposition 2.3.2 we showed that a set can be at best $\sqrt{2}$-good. Let $\eta > 0$ then if $p > 1/\eta^2$ we show below that the set $\alpha_1, \ldots, \alpha_\sigma$ as defined in (2.5.3), is $\sqrt{2-\eta}$-good. Thus the rotations $\alpha_1, \ldots, \alpha_\sigma$ provide sets which are optimally ε-good. To see this note that for f with $\int_{S^2} f \, d\lambda = 0$ and $\|f\| = 1$ we have

$$\sum_{j=1}^{\sigma} \|f_{\alpha_j} - f\|_2^2 = 2\sigma - \langle T_p f, f \rangle \ge p + 1 - 2\sqrt{p}$$

which follows from Theorem 2.5.2. Hence

$$\frac{2}{p+1} \sum_{j=1}^{\sigma} \|f_{\alpha_j} - f\|_2^2 \geq 2 - \frac{4}{\sqrt{p}}.$$

Hence for some $j = 1, \ldots, \sigma$

$$\|f_{\alpha_j} - f\|_2 \geq \sqrt{2 - 4\eta},$$

i.e., $\alpha_1, \ldots, \alpha_\sigma$ is $\sqrt{2 - 4\eta}$-good.

2.6 Distributing points on S^2

The rotations $t_1, t_2, t_3 \in SO(3)$ are clearly ergodic in a very strong sense. The group generated by these is clearly equidistributed in $SO(3)$. More precisely if we consider the $6.5^{\nu-1}$ rotations which are reduced words in t_1, t_2, t_3 (and easily generated recursively) of length ν, they are very evenly distributed in $SO(3)$. This allows us explicitly to generate a large set of rotations which are especially powerful in quadrature. By letting these act on S^2 we get such sequences of points on S^2. In Lubotzky–Phillips–Sarnak [LPS1] an analysis of the distribution of these points on S^2 is examined, and bounds for the discrepancy are determined. In a certain sense (L^2 quadrature) these rotations are optimally equidistributed. Precisely if $W_1, \ldots, W_N \in SO(3)$ are N rotations we define the L^2–discrepancy of the sequence to be

$$D(W_1, \ldots, W_N) = \sup_{\substack{f \in L(G) \\ \|f\|_2 = 1}} \left(\int_G \left| \frac{1}{N} \sum_{j=1}^{N} f(W_j g) - \int_G f(h)\, d\lambda(h) \right|^2 dg \right)^{1/2}.$$

One can show (see Lubotzky–Phillips–Sarnak [LPS1]) that for any choice of N rotations $W_1, \ldots, W_N$ we have

$$D(W_1, \ldots, W_N) \gg N^{-1/2}.$$

On the other hand it follows easily from the considerations in this chapter, (again see [LPS1] for details) that if $N = 6.5^{\nu-1}$ and $U_1, \ldots, U_N$ are the reduced words in t_1, t_2, t_3 of length ν then

$$D(U_1, \ldots, U_N) \ll N^{-1/2} \log N, .$$

We conclude this chapter by noting that we have in fact shown that the only invariant mean on $L^\infty(SO(n))$, $n \geq 3$, is Haar measure. This follows from our analysis since the irreducible representations of $SO(n)$ are precisely the representations above on $H_m(S^{n-1})$. In fact our analysis produces explicit

ε-good sets in $SO(n)$, $n \geq 3$, which are ε-good in the sense that (2.1.1) holds for functions on $G = SO(n)$. The question of the uniqueness of the invariant mean for the general connected compact Lie group is discussed in the notes to this chapter.

We end with a simple question. Given generic $A, B \in SO(n)$, $n \geq 3$, do they form an ε-good set for some $\varepsilon > 0$?

Notes and comments on Chapter 2

Section 2.2: The nonuniqueness of λ for non discrete topological groups which are amenable as discrete groups is due to Granierer [Gr] and Rudin [Ru].

Section 2.3: The reduction Theorem 2.3.1 and its proof are due to Losert and Rindler [LR]. Rosenblatt [Ro] has some related results.

Sections 2.4 and 2.5: The solution of this Ruziewicz problem for $n \geq 4$ is due to Margulis [Ma1] and Sullivan [Su] who used 'property T'. Drinfeld, using adelic automorphic form theory and the Jacquet-Langlands correspondence [JL], proved uniqueness for $n = 2, 3$. The treatment in the text has the advantage of giving a uniform treatment for $n \geq 2$ and more importantly being effective. That is ε-good sets with explicit and effective ε and rotations are constructed. The method using 'property T' has certain advantages. For example Margulis has observed that with the exception of $SO(2)$, $SO(3)$, and $SO(4)$, every simple connected compact Lie group G, has a dense countable subgroup Γ which has 'property T'. This implies (though non-effectively) the existence of an ε-good set in G (for $L^2(G)$), for some $\varepsilon > 0$. It follows from this and the results of this chapter, that every simple connected non-Abelian Lie group has a unique invariant mean. It is easy to see that if G_1 and G_2 have ε-good sets then so does $G_1 \times G_2$. Now it is known that every compact connected Lie group G may be realized in the form

$$G = (T_0 \times G_1 \times \ldots \times G_r)/K \qquad (2.\text{N}.1)$$

where T_0 is the identity component of the center of G, G_i are simply connected simple compact Lie groups and K is a finite subgroup of the center. Hence we conclude that a compact connected Lie group G has a unique invariant mean on $L^\infty(G)$ iff its center is finite.[1]

The treatment of the operators T_p in Section 2.5 uses the theory of Hecke operators on $L^2(S^2)$ developed in Lubotzky-Phillips-Sarnak [LPS1] and follows that paper.

---

[1]We do not know a characterization of this type for the general compact group.

Finally, Chiu [Chi] using the quaternion algebra D/Q which is ramified exactly at 13 and, ∞, has by the techniques of this chapter, constructed an ε-good set in $SO(3)$ with two rotations. He shows that if t_1 is the rotation about the N–S axis through π and t_2 is the rotation through $\cos^{-1}\{1/(2\sqrt{2})\}$ about an axis whose angle with the N–S axis is $\tan^{-1}(\sqrt{13})$ then t_1, t_2 are $\frac{2}{3}(3 - 2\sqrt{2})$-good.

Chapter 3

Ramanujan Graphs

3.1 Counting methods

The thrust of this chapter is the explicit construction of highly connected sparse graphs. Before doing so we first demonstrate the existence of such graphs by elementary counting arguments. We consider two examples; one is the classical problem of the construction of graphs of large girth and large chromatic number (see the introduction for the definitions) the second is the construction of expander graphs.

3.1.1 Large girth and large chromatic number

Our aim is to show that for given integers k and p there is a graph X with girth $\geq k$ and chromatic number $\geq p$. This result is due to Erdős [Er] whose argument runs as follows: Firstly note that any graph X whose maximal independent set (i.e., a subset of the vertices V of X for which no two elements are joined) is of size $i(X)$, satisfies

$$\chi(X) \geq \frac{n}{i(X)}, \qquad (3.1.1)$$

where $\chi(X)$ is the chromatic number. Indeed, if X is colored with r colors then each set A_j, $j = 1, \ldots, r$, of vertices with the same color is clearly independent so that

$$|A_j| \leq i(X)$$

and thus

$$n = \sum_{j=1}^{r} |A_j| \leq i(X)\, r\,.$$

The idea is to construct a graph with large girth and small independence number $i(X)$.

Let k, the girth parameter, be fixed and ≥ 3. We let n be a large integer which will serve as the order, $|X|$, of the graph to be constructed. Also let m be an integer of order $n^{1+\varepsilon}$, where $\varepsilon > 0$ and will be chosen later. m will be the number of edges (which is slightly more than linear). Let p be of order $n^{1-\eta}$ ($\eta > 0$ to be chosen) and it will correspond to the size of the maximal independent set. Consider all graphs X (with labeled vertices) on $V = \{x_1, \ldots, x_n\}$ with m edges. We show that the typical such graph (for n large and X slightly modified) has the desired properties.

The number of labeled graphs X on n vertices with m edges is the number of choices of m edges out of the total of $\binom{n}{2}$ edges, i.e.,

$$\binom{\binom{n}{2}}{m}, \quad X_\alpha\text{'s}, \quad 1 \leq \alpha \leq \binom{\binom{n}{2}}{m}. \tag{3.1.2}$$

Of these we bound from above the number which have at most n edges in every $V^{(p)}$ where $V^{(p)}$ is any subset of X of size p. For a given $V^{(p)}$ the number of graphs as above with ℓ edges in $V^{(p)}$ is

$$\binom{\binom{p}{2}}{\ell} \binom{\binom{n}{2} - \binom{p}{2}}{m - \ell} \tag{3.1.3}$$

and hence the number with at most n edges in $V^{(p)}$ is at most

$$\sum_{\ell=0}^{n} \binom{\binom{p}{2}}{\ell} \binom{\binom{n}{2} - \binom{p}{2}}{m - \ell} < (n+1) \binom{\binom{p}{2}}{m} \binom{\binom{n}{2} - \binom{p}{2}}{m}$$

$$< p^{2n} \binom{\binom{n}{2} - \binom{m}{2}}{m} < \binom{\binom{n}{2}}{m} p^{2n} \left(1 - \frac{\binom{p}{2}}{\binom{n}{2}}\right)^m$$

$$< \binom{\binom{n}{2}}{m} p^{2n} \left(1 - \frac{p^2}{n^2}\right)^m$$

$$< \binom{\binom{n}{2}}{m} p^{2n} e^{-mp^2/n^2}. \tag{3.1.4}$$

The number of choices of $V^{(p)}$ is $\binom{n}{p}$ and so the total number of X_α's with at most n edges from some $V^{(p)}$ is

$$\leq \binom{\binom{n}{2}}{m} p^{3n} e^{-n^{1+\varepsilon-2\eta}} = o\binom{\binom{n}{2}}{m} \tag{3.1.5}$$

if $2\eta < \varepsilon$.

This almost does what we want since clearly there are no independent sets of size p in almost all these graphs. However, the girth as it stands will not be large. We bound the number of X_α's which contain more than n/k

closed circuits of length $\leq k$. We find that the number of these is also $o\left(\binom{\binom{n}{2}}{m}\right)$. Hence for large n we have an X_α (in fact almost every one) satisfying

(i) X_α meets every $V^{(p)}$ in at least $n+1$ edges,

(ii) X_α has $\leq n/k$ closed circuits of length at most k.

If for such an X_α we delete the edges of any closed circuit of length $\leq k$, we will be deleting at most n edges from X_α. Clearly the graph X so obtained has girth $> k$ and also it meets every $V^{(p)}$ in at least one edge. Hence $i(X) < p$ and so $\chi(X_\alpha) \geq n^\eta$. So what remains is to prove the bound for the number of X_α which have more than n/k closed circuits of length k.

A given closed circuit $X_1 \to X_2 \to \ldots \to X_\ell \to X_1$ of length $\ell \leq k$ is in exactly

$$\binom{\binom{n}{2} - \ell}{m - \ell}$$

of the X_α's. Moreover a circuit is considered with order so that there are $n!/(n-\ell)!$ circuits of length ℓ. Hence the expected number of closed circuits of length at most k is

$$\sum_\alpha \text{prob}\,(X_\alpha)\,\#[\text{ closed circuits in } X_\alpha \text{ of length } \leq k]$$

$$= \binom{\binom{n}{2}}{m}^{-1} \sum_{\ell=3}^{k} \frac{n!}{(n-\ell)!} \binom{\binom{n}{2} - \ell}{m - \ell}$$

$$\ll \frac{n^k\,(2m)^k}{n^{2k}} = o(n) \qquad (\text{if } \varepsilon < 1/k \text{ which we assume}).$$

Hence

$$\sum_{\substack{\alpha \\ \text{s.t. \# of closed circuits} \\ \text{of length} \leq k \\ \text{is} > n/k}} \text{prob}\,(X_\alpha)\,\frac{n}{k} = o(n)$$

which implies that almost all X_α (as $n \to \infty$) have at most n/k circuits of length $\leq k$.

Note that the size of the X produced, even with $k = 4$ and $\chi = 4$, is very large.

3.1.2 Expander graphs

Let I and O be two sets of size n (again $n \to \infty$). We want to construct a bipartite graph between I and O (i.e., the edges run between I and O only (see Figure 3.1)) with a linear number of edges, say $k\,n$ of them (k fixed)

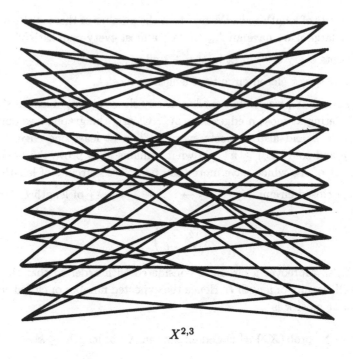

$$X^{2,3}$$

Figure 3.1: Bipartite Graph

and with the following expansion property: For any $A \subset I$ with $|A| \le n/2$ we have

$$|\partial A| \ge c|A|, \qquad\qquad (3.1.6)$$

where $c > 1$ is a constant (clearly $c \le 2$) and $\partial A = \{y \in O | \, (a, y)$ is an edge from some $a \in A\}$.

A bipartite graph satisfying (3.1.6) with the above parameters will be called an (n, k, c) expander. Let us now show by counting arguments that such expanders exist. We do so for $c = 3/2$ and $k = 5$; the general case is similar.

Let $I = O = \{1, 2, \ldots, n\}$ and construct the bipartite graph X by taking k permutations $\pi_1, \ldots, \pi_k$ of I and joining each j to $\pi_r(j)$ for $r = 1, \ldots, k$. This yields a k–regular bipartite graph, i.e., one with each vertex having k edges. The claim is that for $k = 5$ and almost all choices of $\pi = (\pi_1, \ldots, \pi_k)$, X_π satisfies (3.1.6). Clearly there are $(n!)^k$ such π's (though they do not produce distinct X_π's).

Call $\pi = (\pi_1, \ldots, \pi_k)$ bad if for some $A \subset I$ with $|A| \le n/2$ there is $B \subset O$

with $|B| \leq \frac{3}{2}|A|$ for which $\pi_j(A) \subset B$ for $j = 1, \ldots, k$. We want to bound the number of such bad π's. For a given A with $|A| = t \leq n/2$ and B with $t \leq |B| = m \leq \frac{3}{2}t$, the number of bad π's corresponding to this A and B is

$$(m(m-1)\ldots(m-t+1)(n-t)!)^k = \left(\frac{m!(n-t)!}{(m-t)!}\right)^k. \qquad (3.1.7)$$

Hence the total number of bad π's, denoted by BAD, is at most

$$\text{BAD} \leq \sum_{t \leq n/2} \sum_{t \leq m \leq 3t/2} \binom{n}{t}\binom{n}{m}\left(\frac{m!(n-t)!}{(m-t)!}\right)^k$$

$$= \sum_{t \leq n/3} \sum_{t \leq m \leq 3t/2} \binom{n}{t}\binom{n}{m}\left(\frac{m!(n-t)!}{(m-t)!}\right)^k \qquad (3.1.8)$$

$$+ \sum_{n/3 \leq t \leq n/2} \sum_{t \leq m \leq 3t/2} \binom{n}{t}\binom{n}{m}\left(\frac{m!(n-t)!}{(m-t)!}\right)^k$$

$$= I + II$$

Now

$$I \leq n \sum_{t \leq n/3} \binom{n}{t}\binom{n}{3t/2}\left(\frac{(3t/2)!(n-t)!}{(t/2)!}\right)^k$$

$$= n \sum_{t \leq n/3} b_t.$$

For $k \geq 5$, b_t is largest for $t = 1$ so that

$$I \leq n^4((n-1)!)^k, \qquad (3.1.9)$$

and hence $I/(n!)^k \to 0$ as $n \to \infty$. On the other hand

$$II \leq 2^{2n} n \sum_{n/3 \leq t \leq n/2} \left(\frac{(3t/2)!(n-t)!}{(t/2)!}\right)^k = 2^n n \sum_{n/3 \leq t \leq n/2} h_t$$

h_t is largest at one of the endpoints $t = n/3$ or $t = n/2$. Checking at these points in combination with Stirling's formula one finds that $II/(n!)^k \to 0$ as $n \to \infty$. Hence for $k \geq 5$ $\text{BAD}/(n!)^k \to 0$.

Thus expanders as defined in (3.1.6) certainly exist. Surprisingly the explicit construction of graphs satisfying the properties in Sections 3.1.1 and 3.1.2 is rather difficult.

3.2　Spectrum of graphs

Let X be a graph with vertex set $V = \{v_1, \ldots, v_n\}$. The adjacency matrix of X is the $n \times n$ matrix A whose i, j entry a_{ij} is $a_{ij} = \#$ of edges from v_i to v_j. Clearly A is symmetric (our graphs are not directed). From the definition of matrix multiplication we see that if

$$A^r = (a_{ij}^{(r)})$$

then

$$a_{ij}^{(r)} = \# \text{ of paths in } X \text{ of length } r \text{ from } v_i \text{ to } v_j. \qquad (3.2.1)$$

Another linear operator which is closely related to A when X is regular, is the Laplacian Δ defined as follows:

$$\Delta : \ell^2(X) \to \ell^2(X)$$

where $\ell^2(X)$ is the vector space of functions on $V(X)$, by

$$\Delta f(v) = d_v f(v) - \sum_{(w,v) \in E} f(w). \qquad (3.2.2)$$

Here d_v is the degree of the vertex v, that is the number of edges at v. A useful formula for the Laplacian is obtained by integration by parts. Let the edges of X be oriented in an arbitrary fashion. Thus to each e we have vertices e^+ and e^-. Then the following relation holds

$$\sum_{v \in X} f(v) \overline{\Delta g(v)} = \sum_{e \in E} (f(e^+) - f(e^-)) \overline{(g(e^+) - g(e^-))}$$
$$= \sum_e d\,f(e) \,\overline{d\,g(e)} \qquad (3.2.3)$$

with the obvious meaning assigned to $d f$. In particular we see that

$$\langle \Delta f, f \rangle \geq 0. \qquad (3.2.4)$$

Hence Δ is symmetric and nonnegative. It is also clear that $\Delta f = 0$ iff $df = 0$, i.e., f is constant on the connected components of X. Put another way, the spectrum of Δ is real, nonnegative, and the multiplicity of the eigenvalue 0 is the number of connected components of X.

When X is regular of degree k then the matrices A and Δ are related by

$$\Delta = kI - A. \qquad (3.2.5)$$

From now on we will deal with regular graphs only and by $X_{n,k}$ we mean a k–regular graph on n vertices. By the spectrum of $X_{n,k}$ we mean the spectrum of the adjacency matrix A, we denote its eigenvalues by $\lambda_0 \geq \lambda_1 \geq \ldots \geq \lambda_{n-1}$. Clearly $\lambda_0 = k$, corresponding to the constant function and the multiplicity of λ_0 is the number of components of X. It is also obvious that $|\lambda_j| \leq k$.

Lemma 3.2.1. $\lambda_{n-1} = -k$ *iff X is bipartite. In this case the eigenvalues are symmetric about 0.*

Proof: By working on each component of X separately we may assume that X is connected. If $-k$ is an eigenvalue we have

$$\sum_{(w,v)\in E} f(w) = -k\,f(v). \qquad (3.2.6)$$

Hence $|f(v)| \leq \frac{1}{k}\sum_{(w,v)\in E} |f(w)|$, i.e., $|f|$ is 'subharmonic' and hence by the maximum principle it must be constant. f is real so normalizing $f(v) = \pm 1$ we let $I = \{v|\ f(v) = 1\}$ and $O = \{v|\ f(v) = -1\}$. It is clear from (3.2.6) that I and O can have no edges within, i.e., X is bipartite. Conversely if X is bipartite and f an eigenfunction of A then

$$\sum_{(v,w)\in E} f(w) = \lambda\,f(v).$$

By changing the sign of f on one side of the bipartition we get an eigenfunction with eigenvalue $-\lambda$. $\qquad\square$

Definition 3.2.2. We denote by $\mu_1(X_{n,k})$ the absolute value of the next to largest (in absolute value) eigenvalue of $X_{n,k}$. Hence $\lambda_1 \leq \mu_1$.

The point of introducing the spectrum of a graph is that an upper bound on λ_1 ensures that $X_{n,k}$ has the expansion property while an upper bound on μ_1 gives an upper bound on the independence number. In this way we can rephrase the problems introduced earlier in spectral terms.

Proposition 3.2.3.

$$i(X) \leq \frac{\mu_1}{k}\,n.$$

Proof: Let I be an independent subset of vertices with $|I| = r$. Define the function $f(x)$ on X by

$$f(x) = \begin{cases} 1 & \text{if } x \in I \\ -c & \text{if } x \notin I\,, \end{cases}$$

where $r - (n-r)\,c = 0$. Then $f \perp 1$ and so by the variational characterization of the eigenvalues of a symmetric matrix

$$\|A f\|_2^2 \leq \mu_1^2 \|f\|_2^2. \qquad (3.2.7)$$

Since I is independent

$$A f(x) = -ck \qquad \text{for } x \in I.$$

Hence

$$\|A f\|_2^2 \geq c^2 k^2 r. \qquad (3.2.8)$$

Now $c = r/(n - r) = \nu/(1 - \nu)$, where $\nu = r/n$, so combining (3.2.7) and (3.2.8)

$$c^2 k^2 r \leq \mu_1^2 (r + c^2(n - r))$$

$$\Rightarrow \quad c^2 k^2 \leq \mu_1^2 \left(1 + c^2 \left(\frac{1}{\nu} - 1 \right) \right)$$

$$\Rightarrow \quad \nu \leq \frac{\mu_1 n}{k}$$

after some computation, i.e.,

$$|I| \leq \frac{\mu_1 n}{k}. \qquad \qquad \square$$

Thus in order to construct graphs of large chromatic number we need only construct graphs with the ratio μ_1/k small (independent of n).

Next we explain the expander–spectrum connection.

Proposition 3.2.4. *Let $X_{2n,k}$ be a k–regular bipartite graph, then X is an (n, k, c) expander with*

$$c \geq \frac{2 k^2}{k^2 + \lambda_1^2}.$$

Proof: Let $B \subset I$ with $|B| \leq n/2$. Let

$$f(x) = \begin{cases} 1 & \text{if } x \in B \\ 0 & \text{otherwise.} \end{cases}$$

Then

$$A f(y) = \begin{cases} 0 & \text{if } y \notin \partial B \\ c_y, \text{ say,} & \text{if } y \in \partial B. \end{cases}$$

The number of edges running between B and ∂B is, on the one hand $k|B|$, while on the other it can be expressed as $\sum_{y \in \partial B} c_y$. Hence

$$\sum_{y \in \partial B} c_y = k|B|. \qquad (3.2.8a)$$

Now

$$\sum_x |(A f)(x)|^2 = \sum_{y \in \partial B} |c_y|^2 \geq \left(\sum_{y \in \partial B} c_y \right)^2 \frac{1}{|\partial B|},$$

hence

$$\sum_x |A f(x)|^2 \geq \frac{(k|B|)^2}{|\partial B|}. \tag{3.2.8b}$$

Let ϕ_0^+ and ϕ_0^- be the normalized eigenfunctions of A corresponding to the eigenvalues k and $-k$ respectively. We may write

$$f(x) = \frac{|B|}{\sqrt{2n}} \phi_0^+(x) + \frac{|B|}{\sqrt{2n}} \phi_0^-(x) + h(x), \tag{3.2.8c}$$

with $\langle h, \phi_0 \rangle = \langle h, \phi_1 \rangle = 0$. Then

$$\|A f\|^2 \leq \frac{k^2 |B|^2}{n} + \lambda_1 \langle h, h \rangle. \tag{3.2.8d}$$

Also

$$\langle h, h \rangle = \langle f, f \rangle - \frac{|B|^2}{n} = |B| - \frac{|B|^2}{n}.$$

Combining this with (3.2.8b)

$$\frac{k^2 |B|^2}{|\partial B|} \leq \frac{k^2 |B|^2}{n} + \lambda_1 \left(|B| - \frac{|B|^2}{n} \right)$$

so

$$\frac{k^2 |B|^2}{|\partial B|} \leq (k^2 - \lambda_1^2) \frac{|B|}{n} + \lambda_1^2,$$

$$\leq (k^2 - \lambda_1^2) \frac{1}{2} + \lambda_1^2 = \frac{k^2 + \lambda_1^2}{2},$$

i.e.,

$$|\partial B| > \frac{2k^2}{k^2 + \lambda_1^2}. \qquad \qquad \square$$

By the same argument one can prove:

Proposition 3.2.5. *Let $X_{n,k}$ be an arbitary k–regular graph; then for any set $B \subset V$ with $|B| \leq n/2$*

$$|\partial B| \geq c |B|$$

with $c = (k - \lambda_1)/2k$. (Such graphs have been called (n, k, c) enlargers by Alon [A1].)

Also along the same theme we can bound the diameter of an $X_{n,k}$ in terms of μ_1.

Proposition 3.2.6.

$$\operatorname{diam}(X_{n,k}) \leq \frac{\log(2n)}{\log\left(\frac{k + \sqrt{k^2 - \mu_1^2}}{\mu_1}\right)}.$$

Proof: Let $\phi_j(v)$, $j = 0, 1, \ldots, n-1$ be an orthonormal basis of $\ell^2(V)$ consisting of eigenfunctions of A, with eigenvalues λ_j. Our assumption is that $|\lambda_j| \leq \mu_1$ for $j \neq 0$. Also $\phi_0(v) = 1/\sqrt{n}$, $\lambda_0 = k$. For any polynomial P we have, on applying the spectral theorem,

$$P(A)(v,w) = \sum_{j=0}^{n-1} P(\lambda_j)\,\phi_j(v)\,\phi_j(w). \qquad (3.2.9)$$

For $v, w \in V$ suppose $\text{dist}(v,w) > N$, then clearly

$$P(A)(v,w) = 0 \qquad (3.2.10)$$

for any polynomial of degree N. Using this in (3.2.9) yields

$$0 = \sum_{j=0}^{n-1} P(\lambda_j)\,\phi_j(v)\,\phi_j(w)$$

or

$$\frac{P(k)}{n} = -\sum_{j=1}^{n-1} P(\lambda_j)\phi_j(v)\,\phi_j(w)$$

$$\leq \sup_{|\lambda|\leq\mu_1} |P(\lambda)| \sum_{j=1}^{n-1} |\phi_j(v)\,\phi_j(w)|$$

$$\leq \sup_{|\lambda|\leq\mu_1} |P(\lambda)| \sum_{j=1}^{n-1} \frac{1}{2}\left(|\phi_j(v)|^2 + |\phi_j(w)|^2\right)$$

$$\leq \sup_{|\lambda|\leq\mu_1} |P(\lambda)|,$$

i.e., we have, for any polynomial of degree $\leq N$

$$P(k) \leq n \sup_{|\lambda|\leq\mu_1} |P(\lambda)|. \qquad (3.2.11)$$

To get the maximum out of (3.2.11) we apply it with $P(x)$, a Chebyshev polynomial, since these have well-known extremal properties in L^∞. Let

$$P_N(x) = T_N(x/\mu_1),$$

where

$$T_N(x) = \cos(N \arccos(x)) = \frac{1}{2}\left\{(x + i\sqrt{1-x^2})^N + (x - i\sqrt{1-x^2})^N\right\}.$$

Clearly $P_N(x) \leq 1$ for $|x| \leq \mu_1$ hence we get from (3.2.11) that

$$P_N(k) \leq n$$

and so

$$\frac{1}{2}\left(\frac{k+\sqrt{k^2-\mu_1^2}}{\mu_1}\right)^N \leq n$$

or

$$N \leq \frac{\log 2n}{\log\left(\frac{k+\sqrt{k^2-\mu_1^2}}{\mu_1}\right)}.$$

The last three Propositions show that in order to obtain the desired properties for an $X_{n,k}$ it is desirable to make $\lambda_1(X)$ and $\mu_1(X)$ as small as possible. This leads us to the definition of a Ramanujan graph. Before giving it we prove a Proposition which limits from below the size of λ_1 (and hence μ_1).

Proposition 3.2.7. *Let k be fixed, then*

$$\lim_{n\to\infty}\left(\inf_{|X_{m,k}|\geq n}\lambda_1(X_{m,k})\right) \geq 2\sqrt{k-1},$$

the inf is taken over all $X_{m,k}$'s with $\geq n$ vertices.

Proof: We must show that if $\varepsilon > 0$ is given then for n large enough $\lambda_1(X_{n,k}) \geq 2\sqrt{k-1}-\varepsilon$ for any $X_{n,k}$. Suppose not, that is there are arbitrary large n's and corresponding $X_{n,k}$'s for which

$$\lambda_1(X_{n,k}) \leq 2\sqrt{k-1} - \varepsilon. \tag{3.2.12}$$

Consider for $r \geq 0$

$$\mathrm{tr}\,(A^r) = \sum_{j=1}^{n} a_{jj}^{(r)} = \sum_{j=0}^{n-1} \lambda_j^r. \tag{3.2.13}$$

Now $a_{jj}^{(r)}$ is the number of paths of length r from v_j to v_j. Clearly this number is greater than or equal to the number of paths of length r from v to v, where v is any vertex of the infinite k–regular tree. This may be seen by realizing the k–regular tree as the universal covering of $X_{n,k}$ and noting that every closed path from v to v on the tree gives rise to one on $X_{n,k}$ (here v is a lift of v_j). Let $\rho(r)$ denote the number of paths of length r on the tree from some vertex v to itself. We have

$$\sum_{j=0}^{n-1} \lambda_j^r \geq n\,\rho(r). \tag{3.2.14}$$

$\rho(r)$ may easily be computed (we leave it to the reader):

$$\rho(r) = 0 \quad \text{for } r \text{ odd},$$

$$\rho(2m) \geq \frac{1}{m}\binom{2m-2}{m-1} k\,(k-1)^{m-1}. \tag{3.2.15}$$

Hence

$$\frac{1}{n} \sum_{j=0}^{n-1} \lambda_j^{2m} \geq \rho(2m),$$

and using (3.2.12) we get

$$\frac{k^{2m}}{n} + \frac{1}{n} \sum_{|\lambda_j| \leq 2\sqrt{k-1}-\varepsilon} \lambda_j^{2m} + \frac{1}{n} \sum_{-k \leq \lambda_j < -2\sqrt{k-1}+\varepsilon} \lambda_j^{2m} \geq \rho(2m) \tag{3.2.15}$$

$$I + II + III \geq \rho(2m)$$

Now $(\rho(2m))^{1/(2m)} \to 2\sqrt{k-1}$ as $m \to \infty$, so choosing m large enough we can arrange

$$\rho(2m)^{1/(2m)} > 2\sqrt{k-1} - \varepsilon/2. \tag{3.2.16}$$

Also

$$II \leq (2\sqrt{k-1} - \varepsilon)^{2m}$$

so for m large enough, say $m \geq M$,

$$\frac{k^{2m}}{n} + \frac{1}{n} \sum_{\lambda_j < -2\sqrt{k-1}+\varepsilon} \lambda_j^{2m} \geq \frac{1}{2} (2\sqrt{k-1} - \varepsilon/2)^{2m}. \tag{3.2.17}$$

(Here M depends only on k and ε.)

Hence for n large enough

$$\frac{1}{n} \sum_{\lambda_j < -2\sqrt{k-1}+\varepsilon} \lambda_j^{2m} \geq \frac{1}{3} (2\sqrt{k-1} - \varepsilon/2)^{2m}. \tag{3.2.18}$$

However we also have from (3.2.15)

$$\frac{1}{n} \sum_{\lambda_j < -2\sqrt{k-1}+\varepsilon} \lambda_j^{2m+1} + \frac{1}{n} k^{2m+1} + O((2\sqrt{k-1} \mp \varepsilon)^{2m+1}) \geq 0.$$

If we substitute this into (3.2.18) we get

$$-\frac{1}{3} (2\sqrt{k-1} - \varepsilon/2)^{2m} (2\sqrt{k-1} - \varepsilon) + \frac{1}{n} k^{2m+1} + (2\sqrt{k-1} - \varepsilon)^{2m+1} \geq 0.$$

For n large enough this is impossible.   □

Finally we make the definition

Definition 3.2.8. A graph $X_{n,k}$ is called *Ramanujan* if

$$\mu_1(X_{n,k}) \leq 2\sqrt{k-1}.$$

A bipartite $X_{n,k}$ is called a *bipartite Ramanujan graph* if

$$\lambda_1(X_{n,k}) \leq 2\sqrt{k-1}.$$

Proposition 3.2.7 asserts that asymptotically as $n \to \infty$ Ramanujan graphs are optimal in minimizing λ_1 and μ_1. Of course they enjoy in a strong way the properties ensured by Propositions 3.2.3, 3.2.5, and 3.2.6. For small values of n and k one can check directly that a graph is Ramanujan. That they exist with $n \to \infty$ is the subject matter of the next section.

3.3 Explicit Ramanujan graphs

The Kloosterman sum bound of Weil (1.5.8) shows that the numbers $\bar{x} \pmod{p}$ as x runs through $1, 2, \ldots, p-1$ come down in a 'random' manner, that is the cancellation of the p terms in the series (each of which is a root of 1) is as much as $\sqrt{p}$. This is what one would expect of a random sequence. Hence we may expect that the graph X defined below through such arithmetic operations, to have expansion properties.

Let the vertices of X_p be the points $\{0, \ldots, p-1, \infty\}$ of $P^1(\mathbf{F}_p)$, the projective line over the field with p elements. Join $\xi \to \xi + 1$, $\xi - 1$ and $\bar{\xi}$ for each $\xi \in P^1(\mathbf{F}_p)$. In this way we get a 3–regular graph with $p+1$ vertices. We might expect that X_p is an expanding family. Actually one can show that $\lambda_1(X_p) \leq 2.999$[1] and so by the results of the last Section X_p does have the desired properties. The bound however is poor and we now give an explicit family of Ramanujan graphs.

We begin with the description of the graphs. Let p, q be unequal primes both $\equiv 1 \pmod{4}$. This last restriction can easily be removed; we have imposed it simply to keep the description simple. Let i be an integer satisfying $i^2 \equiv -1 \pmod{q}$. From (1.1.6) we know that there are $8(p+1)$ solutions $\alpha = (a_0, a_1, a_2, a_3)$ to

$$a_0^2 + a_1^2 + a_2^2 + a_3^2 = p. \tag{3.3.1}$$

Among these there are exactly $p+1$ with $a_0 > 0$ and odd and a_j, $j = 1, 2, 3$ even. To each such α associate the matrix $\tilde{\alpha}$ in $PGL(2, \mathbf{Z}/q\mathbf{Z})$, by

$$\tilde{\alpha} = \begin{pmatrix} a_0 + i\,a_0 & a_2 + i\,a_3 \\ -a_2 + i\,a_3 & a_0 - i\,a_1 \end{pmatrix}. \tag{3.3.2}$$

This gives us $k = p+1$ matrices in $PGL(2, \mathbf{Z}/q\mathbf{Z})$.

---

[1]For $k = 3$ the Ramanujan graph has $\lambda_1 \leq 2\sqrt{2} = 2.828\ldots$. For a proof of the 2.999 bound see [LPS2].

Our graphs will be Cayley graphs of the group $PGL(2, \mathbf{Z}/q\mathbf{Z})$ relative to the above generators. Quite generally if G is a group and S a set of generators of G which is moreover symmetric (i.e., $s \in S \Rightarrow s^{-1} \in S$), then one can construct a graph called the Cayley graph of G relative to S as follows: The vertices of X are the elements of G while the edges run from g to sg for each $s \in S$. In this way we get a $|S|$-regular graph on $|G|$ vertices.

Returning to our case of $G = PGL(2, \mathbf{Z}/q\mathbf{Z})$ and S the set of $p+1$ elements of G as above (S is symmetric) we get a graph of order $n = q(q^2 - 1)$. If $\left(\frac{p}{q}\right) = 1$ then this graph is not connected since the elements of S all lie in the index two subgroup $PSL(2, \mathbf{Z}/q\mathbf{Z})$ (i.e., elements whose determinant is a square). We define the Cayley graphs $X^{p,q}$ to be the above Cayley graph if $\left(\frac{p}{q}\right) = -1$ and to be the Cayley graph of $PSL(2, \mathbf{Z}/q\mathbf{Z})$ relative to S if $\left(\frac{p}{q}\right) = 1$. The graphs $X^{p,q}$ will be shown to be connected. If $\left(\frac{p}{q}\right) = -1$, $X^{p,q}$ is bipartite, the bipartition corresponding to the subgroup $PSL(2, \mathbf{Z}/q\mathbf{Z})$ and its complement. Thus $X^{p,q}$ is a $k = (p+1)$-regular graph on $n = q(q^2 - 1)$ or $q(q^2 - 1)/2$ vertices depending on the sign of $\left(\frac{p}{q}\right)$.

Theorem 3.3.1.

Case (i): $\left(\frac{p}{q}\right) = -1$,

(a) $X^{p,q}$ is a bipartite Ramanujan graph,

(b) girth $(X^{p,q}) \geq 4 \log_p q - \log_p 4$,

(c) diam $(X^{p,q}) \leq 2 \log n + 2 \log_p 2 + 1$.

Case (ii): $\left(\frac{p}{q}\right) = 1$,

(a) $X^{p,q}$ is a Ramanujan graph,

(b) girth $(X^{p,q}) \geq 2 \log_p q$,

(c) diam $(X^{p,q}) \leq 2 \log_p n + 2 \log_p 2 + 1$.

(d) $i(X^{p,q}) \leq \dfrac{2\sqrt{p}}{p+1} n$,

(e) $\chi(X^{p,q}) \geq \dfrac{p+1}{2\sqrt{p}}$.

In particular $X^{p,q}$, $\left(\frac{p}{q}\right) = 1$ give explicit graphs of large girth and small independence number and hence have also large chromatic number. Also in view of Proposition 3.2.4 these graphs with $\left(\frac{p}{q}\right) = -1$ give explicit expanders with coefficient of expansion as in that Proposition. The statements (i)(c), (ii)(c), (d), and (e) all follow from the Ramanujan property (a) and

Propositions 3.2.3 and 3.2.6. One comment here is that as it stands Proposition 3.2.6 does not apply to the bipartite Ramanujan case, however the proof given there is easily modified to include this as well. So what needs to be shown is that $X^{p,q}$ is Ramanujan (or bipartite Ramanujan) as well as establishing the girth lower bounds. Concerning these we note that for $\left(\frac{p}{q}\right) = -1$ the $X^{p,q}$ give k-regular graphs of order n with

$$\text{girth}\,(X_{n,k}) \geq \frac{4}{3}\,\log_{k-1} n\,. \qquad (3.3.3)$$

This in fact is the (asymptotically) largest known girth for k-regular graphs. Random (or counting methods) graphs have girths $g(X_{n,k})$ asymptotically $\log_{k-1} n$ [Bo2]. It is also clear on the other hand that $\text{girth}\,(X_{n,k}) \leq 2\log_{k-1} n$ for any $X_{n,k}$.

Concerning the diameter of graphs $X^{p,q}$ we expect that it is essentially as small as possible. Precisely, for $\varepsilon > 0$

$$\text{diam}\,(X^{p,q}) \leq (1 + \varepsilon)\log n + C_\varepsilon$$

as $n \to \infty$.

3.4 Proofs

In this section we give proofs of the claims made in Theorem 3.3.1. As in Chapter 2 the integral quaternions $H(\mathbf{Z})$ play a central role and we will use the same notation as in Chapter 2. Let

$$\Lambda'(2) = \{\alpha \in H(\mathbf{Z})| \; \alpha \equiv 1(\text{mod}\,2) \text{ and } N(\alpha) = p^\nu, \; \nu \in \mathbf{Z}\}\,.$$

Here p is our fixed prime $p \equiv 1 \pmod 4$. $\Lambda'(2)$ is closed under multiplication and if we identify α and β in $\Lambda'(2)$ whenever $\pm p^{\nu_1}\alpha = p^{\nu_2}\beta$, $\nu_1, \nu_2 \in \mathbf{Z}$ then the equivalence classes so obtained form a group with $[\alpha][\beta] = [\alpha\beta]$ and $[\alpha][\bar{\alpha}] = [1]$. Lemma 2.5.4 implies that this group which we denote by $\Lambda(2)$, is free on $[\alpha_1], \ldots, [\alpha_s]$. The Cayley graph of $\Lambda(2)$ with respect to the set S is therefore a tree of degree $p+1$. This tree will be denoted by $\Lambda(2)$ as well. We have thus realized this free group or tree in a suitable number theoretic way. In order to form finite graphs we choose a normal subgroup Γ of $\Lambda(2)$ of finite index. Then Γ acts on $\Lambda(2)$ by multiplication on the right and the quotient graph (or group) $\Lambda(2)/\Gamma$ is finite. This is then the Cayley graph of $\Lambda(2)/\Gamma$ with respect to the generators $\alpha_1\Gamma, \alpha_2\Gamma, \ldots, \overline{\alpha_s}\,\Gamma$.

In order to have any number theoretic significance we must choose Γ in an appropriate way. Let $(m, p) = 1$ and consider all $[\alpha] \in \Lambda(2)$ such that $2m|a_j$, $j = 1, 2, 3$ where $\alpha = a_0 + a_1\mathbf{i} + a_2\mathbf{j} + a_3\mathbf{k}$. This defines a subgroup

$\Lambda(2m)$ of $\Lambda(2)$. It is a normal subgroup of finite index in $\Lambda(2)$ since it may be viewed as follows.

Let $H(\mathbf{Z}/(2m)\mathbf{Z})$ be the ring of quaternions with entries in $\mathbf{Z}/(2m)\mathbf{Z}$ and $H(\mathbf{Z}/(2m)\mathbf{Z})^*$ the invertible elements in this ring. Let $Z \subseteq H(\mathbf{Z}/(2m)\mathbf{Z})^*$ be the central subgroup $Z = \{a_0 : a_0 \neq 0\}$. The homomorphism $\phi : \Lambda(2) \to H(\mathbf{Z}/(2m)\mathbf{Z})^*/Z$ defined by $[\alpha] \to (\alpha \bmod 2m)\, Z$ is well defined. Its kernel is $\Lambda(2m)$.

Now let $m = q$ as in Section 3.3. We show that the graph $X^{p,q}$ may be identified with the Cayley graph of the group $\Lambda(2)/\Lambda(2q)$ with respect to the generators $\alpha_1, \ldots, \overline{\alpha_s}$. This will establish that $X^{p,q}$ is connected.

Define the homomorphism $\phi : \Lambda(2) \to PGL(2, \mathbf{Z}/q\mathbf{Z})$ by

$$[\alpha] \quad \begin{array}{c} \xrightarrow{\pi} \\ \xrightarrow{\phi} \end{array} \alpha \bmod q \xrightarrow{\sigma} \begin{pmatrix} a_0 + i\, a_1 & a_2 + i\, a_3 \\ -a_2 + i\, a_3 & a_0 - i\, a_1 \end{pmatrix}$$

where i is a fixed integer satisfying $i^2 \equiv -1 \pmod q$.

Proposition 3.4.1.

$$\text{Image } \phi = \begin{cases} PGL(2, \mathbf{Z}/q\mathbf{Z}) & \left(\frac{p}{q}\right) = -1 \\ PSL(2, \mathbf{Z}/q\mathbf{Z}) & \left(\frac{p}{q}\right) = 1 \end{cases}$$

Proof: If $\alpha_j \in H(\mathbf{Z})$ is of norm p then $\phi(\alpha_j)$ is in $PSL(2, \mathbf{Z}/q\mathbf{Z})$ iff $\left(\frac{p}{q}\right) = 1$. Since $[PGL(2, \mathbf{Z}/q\mathbf{Z}); PSL(2, \mathbf{Z}/q\mathbf{Z})] = 2$, it suffices to show that $\phi(\Lambda(2)) \supseteq PSL(2, \mathbf{Z}/q\mathbf{Z})$. Now ϕ factors as

$$\Lambda(2) \xrightarrow{\pi_1} H(\mathbf{Z}/(2q)\mathbf{Z})^*/Z \xrightarrow{\pi_2} H(\mathbf{Z}/q\mathbf{Z})^*/Z \xrightarrow{\pi_3} PGL(2, \mathbf{Z}/q\mathbf{Z}) \,.$$

π_3 is an isomorphism so what needs to be checked is the image of $\pi_2 \circ \pi_1$. To prove the Proposition it suffices to show that if $\beta = b_0 + b_1\mathbf{i} + b_2\mathbf{j} + b_3\mathbf{k}$ is in $H(\mathbf{Z}/q\mathbf{Z})$ and is of norm $\equiv 1 \pmod q$ then there is an $\alpha \in H(\mathbf{Z})$ with $N(\alpha) = p^k$, $\alpha \equiv 1 \pmod 2$ and $\alpha \equiv \beta \pmod q$. Let such a β be given. Set $\gamma = \gamma_0 + \gamma_1\mathbf{i} + \gamma_2\mathbf{j} + \gamma_3\mathbf{k}$ where $\gamma_0 \equiv b_0 \pmod q$, $2\,\gamma_j \equiv b_j \pmod q$, $j = 1, 2, 3$. Then

$$\gamma_0^2 + 4\gamma_1^2 + 4\gamma_2^2 + 4\gamma_3^2 \equiv 1 \pmod q \,.$$

We need some results from the theory of quadratic Diophantine equations and in particular the singular series of Hardy and Littlewood. Malyzev [Mal] obtained the following: Let $f(x_1, \ldots, x_n)$ be a quadratic form in $n \geq 4$ variables with integral coefficients and discriminant α. Let $(g, 2d) = 1$ be such that for m sufficiently large with $(g, 2md) = 1$ and m generic for f (i.e., $f = m$ may be solved mod ℓ for every ℓ) and if $(b_1, \ldots, b_n, g) = 1$, $f(b_1, \ldots, b_n) \equiv$

$m \pmod{g}$, then there are integers $(a_1, \ldots, a_n) \equiv (b_1, \ldots, b_n) \pmod{g}$ such that $f(a_1, \ldots, a_n) = m$. Indeed he obtained an asymptotic formula for the number of such $(a_1, \ldots, a_n)$ as $m \to \infty$ (i.e., the singular series).

We apply this to

$$f(x_1, x_2, x_3, x_4) = x_1^2 + 4x_2^2 + 4x_3^2 + 4x_4^2,$$

$m = p^k$, $g = q$, and $(b_0, b_1, b_2, b_3) = (\gamma_0, \gamma_1, \gamma_2, \gamma_3)$. If k is large enough and $p^k \equiv 1 \pmod{q}$ then $f(\gamma_0, \gamma_1, \gamma_2, \gamma_3) \equiv p^k \pmod{q}$ and p^k is generic for f. Hence there is an $(a_0, a_1, a_2, a_3) \equiv (\gamma_0, \gamma_1, \gamma_2, \gamma_3) \pmod{q}$ satisfying

$$a_0^2 + 4a_1^2 + 4a_2^2 + 4a_3^2 = p^k.$$

Hence if $\alpha = a_0 + 2a_1 \mathbf{i} + 2a_2 \mathbf{j} + 2a_3 \mathbf{k}$ then $N(\alpha) = p^k$, $\alpha \equiv 1 \pmod{2}$, and $\alpha \equiv \beta \pmod{q}$.

From Proposition 3.4.1 it follows that $\Lambda(2)/\Lambda(2q) \cong PGL(2, \mathbf{Z}/q\mathbf{Z})$ or $PSL(2, \mathbf{Z}/q\mathbf{Z})$ depending on the sign of $\left(\frac{p}{q}\right)$. Furthermore the homomorphism takes the generators $\alpha_1, \ldots, \overline{\alpha_s}$ to the matrices (3.3.2) and hence $X^{p,q}$ may be identified with $\Lambda(2)/\Lambda(2q)$. Next we prove the lower bound for the girth, i.e., part (b) of Theorem 3.3.1.

3.4.1   Girth lower bound

$X^{p,q}$ is a Cayley graph and hence homogeneous. The shortest circuit may therefore be assumed to run from the identity to itself. On the tree $\Lambda(2)$ this amounts to the length of the smallest member of $\Lambda(2q)$. If $\gamma \in \Lambda(2q)$, $\gamma \neq e$, is of length t then we can find an integral quaternion $\tilde{\gamma} \in \Lambda'(2)$ such that

$$\tilde{\gamma} = \beta_1 \beta_2 \ldots \beta_t \qquad \text{with } \beta_j \in \{\alpha_1, \ldots, \overline{\alpha_s}\}$$

and $\tilde{\gamma} \in \Lambda'(2q)$. Thus $N(\tilde{\gamma}) = p^t$ and $\tilde{\gamma} = a_0 + 2qa_1 \mathbf{i} + 2qa_2 \mathbf{j} + 2qa_3 \mathbf{k}$, $a_\nu \in \mathbf{Z}$. Now as $\gamma \neq e$ at least one of a_1, a_2, a_3 is nonzero. Thus we have

$$p^t = a_0^2 + 4q^2 a_1^2 + 4q^2 a_2^2 + 4q^2 a_3^2. \tag{3.4.1}$$

In case $\left(\frac{p}{q}\right) = 1$ we observe that since some $0 \neq a_j$ for some $j = 1, 2, 3$,

$$p^t \geq 4q^2$$

or $t \geq 2 \log_p q$ as claimed.

In case $\left(\frac{p}{q}\right) = -1$ we first note that t must be even, for if not we would have on reducing mod q.

$$\left(\frac{p}{q}\right)^t = 1 \qquad \text{or} \qquad \left(\frac{p}{q}\right) = 1.$$

Thus $t = 2r$ say. In this case (3.4.1) has the trivial solutions $a_0 = \pm p^r$. The congruence

$$X_0^2 \equiv p^t \pmod{q^2} \tag{3.4.2}$$

has as its only solutions

$$X_0 = \pm p^r \pmod{q^2}.$$

If we assume that (3.4.1) has a non-trivial solution with

$$p^t < \frac{q^4}{4} \tag{$*$}$$

then $p^r < q^2/2$ and so any solution X_0 of (3.4.2) which is not $\pm p^r$ will by the above satisfy

$$|X_0| \geq \frac{q^2}{2}.$$

Hence $X_0^2 \geq q^4/4$. But then clearly $p^t > q^4/4$ contradicting $(*)$. Thus we have $p^t > q^4/4$ or

$$t \geq \frac{4 \log q - \log 4}{\log p}.$$

Finally note that when $\left(\frac{p}{q} \right) = 1$, $X^{p,q}$ is not bipartite (this will be used later on). For if it were we would have $X = PSL(2, \mathbf{Z}/q\mathbf{Z})$ partitions into A and B such that $\alpha_j A = B$ and $\alpha_j B = A$ for each $\alpha_1, \ldots, \overline{\alpha_s}$. Clearly then A is a subgroup (in fact the subgroup of elements expressible as a product of an even number of elements of $\alpha_1, \ldots, \overline{\alpha_s}$) of $PSL(2, \mathbf{Z}/q\mathbf{Z})$, of index two. Hence A is normal, but since for $q > 3$, $PSL(2, \mathbf{Z}/q\mathbf{Z})$ is simple, this is impossible.

To complete the proof of Theorem 3.3.1 we still need to verify that $X^{p,q}$ is Ramanujan.

3.5   Proof of Theorem 3.3.1

One may view the spectral theory of A on $X = \Lambda(2)/\Gamma = T/\Gamma$, where T is the $p+1$ regular tree and Γ a discontinuous group of automorphisms acting on T, as the spectral analysis of A acting on Γ periodic (automorphic) functions on T. First we consider the operator A acting on all functions on T;

$$A f(x) = \sum_{d(y,x)=1} f(y), \qquad d = \text{distance on the tree.} \tag{3.5.1}$$

The kernel of A, call it $k_A(x,y)$, takes the form

$$k_A(x,y) = \begin{cases} 1 & \text{if } d(x,y) = 1 \\ 0 & \text{if } d(x,y) \neq 1. \end{cases}$$

Let A_n be the operator with kernel $k_n(x,y)$ where

$$k_n(x,y) = \begin{cases} 1 & \text{if } d(x,y) = n \\ 0 & \text{otherwise.} \end{cases}$$

i.e.,

$$A_n f(x) = \sum_{d(x,y)=n} f(y).$$

The A_n's are in the polynomial algebra generated by A; in fact a moment's thought shows for $n \geq 2$

$$\sum_y k_n(x,y)\, k_1(y,z) = k_{n+1}(x,z) + p\, k_{n-1}(x,z)$$

and so

$$A_n A_1 = A_1 A_n = A_{n+1} + p A_{n-1}. \tag{3.5.2}$$

A straightforward calculation with (3.5.2) then shows that

$$\sum_{0 \leq r \leq t/2} A_{t-2r} = p^{t/2} U_t\left(\frac{A_1}{2\sqrt{p}}\right) \tag{3.5.3}$$

where U_t is the Chebyshev polynomial of the second kind

$$U_t(x) = \frac{\sin((t+1)\arccos x)}{\sin(\arccos x)}. \tag{3.5.4}$$

We now restrict the operators A_m to $\Lambda(2q)$ automorphic functions; i.e., to the finite dimensional space $X^{p,q} = \Lambda(2)/\Lambda(2q)$ and compute the trace of the operators in (3.5.3). Let the spectrum of $\Lambda(2)/\Lambda(2q)$ as a graph (or viewed as the spectrum of A_1 on $\Lambda(2q)$ automorphic forms!) be as usual $\lambda_0 \geq \lambda_1 \geq \ldots \geq \lambda_{n-1}$. Write

$$\lambda_j = 2\sqrt{p} \cos\theta_j. \tag{3.5.5}$$

Then

$$\text{trace}\left(p^t U_t\left(\frac{A_1}{2\sqrt{p}}\right)\right) = \sum_{j=0}^{n-1} p^{t/2} \frac{\sin(t+1)\theta_j}{\sin\theta_j}. \tag{3.5.6}$$

On the other hand we can compute the trace from the right hand side of (3.5.3), i.e., geometrically

$$\text{trace}\, A_{t-2\sigma}\Big|_{\ell^2(\Lambda(2)/\Lambda(2q))} = \sum_{x \in \Lambda(2)/\Lambda(2q)} \sum_{\gamma \in \Lambda(2q)} k_{t-2r}(x, \gamma(x)).$$

Since $\Lambda(2q)$ is a normal subgroup of $\Lambda(2)$ we have that $\Lambda(2)/\Lambda(2q)$ is a homogeneous graph so that for any x

$$\sum_{\gamma\in\Lambda(2q)} k_{t-2r}(x,\gamma x) = \sum_{\gamma\in\Lambda(2q)} k_{t-2r}(e,\gamma e).$$

Hence

$$\text{trace } A_{t-2r}\Big|_{\ell^2(X)} = |X|\sum_{\gamma\in\Lambda(2q)} k_{t-2r}(e,\gamma e)$$

$$= |X|\,|\{\gamma\in\Lambda(2q)|\ d(\gamma e,e) = t-2r\}|. \qquad (3.5.7)$$

Let

$$Q(x_1,x_2,x_3,x_4) = x_1^2 + (2q)^2 x_2^2 + (2q)^2 x_3^2 + (2q)^2 x_4^2. \qquad (3.5.8)$$

As in Chapter 1 let $r_Q(\nu)$ be the number of representations of ν by Q. Clearly $r_Q(p^k)$ is the number of $\alpha\in H(\mathbf{Z})$ such that $2q|\alpha-a_0$ and $N(\alpha) = p^k$. By Lemma 2.5.4 every such α is uniquely of the form $\pm p^r R_t(\alpha_1,\dots,\overline{\alpha_s})$ where $2r+t = k$ and where $[\alpha]\in\Lambda(2q)$. It follows that

$$r_Q(p^k) = 2\sum_{r\le k/2} |\{\alpha\in\Lambda(2q)|\ d(\alpha,e) = k-2r\}|. \qquad (3.5.9)$$

We have used the fact that reduced word length in the α's corresponds to distance on the tree $\Lambda(2)$. Combining (3.5.9) with (3.5.7) and (3.5.6) we get the identity

$$r_Q(p^k) = \frac{2p^{k/2}}{n}\sum_{j=0}^{n-1}\frac{\sin(k+1)\theta_j}{\sin\theta_j} \qquad (3.5.10)$$

We are now in the position to apply the theory developed in Chapter 1, indeed we are precisely in the Ramanujan type setup. From (1.3.12) and (1.4.10) we have

$$r_Q(p^k) = \delta(p^k) + a(p^k) \qquad (3.5.11)$$

where $a(p^k)$ is the Fourier coefficient of a cusp form of weight two for $\Gamma(16\,q^2)$ while $\delta(p^k)$ is the coefficient of an Eisenstein series of weight two. In view of the calculations made in Section 1.4 we know that δ is of the form

$$\delta(m) = \sum_{d|m} d\,F(d) \qquad (3.5.12)$$

where $F : \mathbf{N}\to\mathbf{C}$ is periodic of period $16\,q^2$.

Lemma 3.5.1. Let $G : \mathbf{N}\to\mathbf{C}$ be periodic and satisfy

$$\sum_{d|p^k} d\,G(d) = o(p^k) \qquad \text{as } k\to\infty$$

then

$$\sum_{d|p^k} d\, G(d) = 0 \qquad \text{for all } k.$$

Proof: Let $\alpha_k = \sum_{d|p^k} d\, G(d)$, then

$$\frac{\alpha_k}{p^k} - \frac{\alpha_{k-1}}{p^{k-1} p} = G(p^k).$$

Now as $k \to \infty$ the left-hand side $\to 0$ and since G is periodic it follows that $G(p^k) = 0$ for all k. $\qquad\square$

Returning to (3.5.10) and (3.5.11) we have

$$\delta(p^k) + a(p^k) = \frac{2\, p^{k/2}}{n} \sum_{j=0}^{n-1} \frac{\sin(k+1)\theta_j}{\sin\theta_j}.$$

The term $a(p^k)$ may be estimated by the Ramanujan bound

$$a(p^k) = O_\epsilon(p^{k(1/2+\epsilon)}). \tag{3.5.13}$$

We have

$$\delta(p^k) + O_\epsilon(p^{k(1/2+\epsilon)}) = \frac{2\, p^{k/2}}{n} \sum_{j=0}^{n-1} \frac{\sin(k+1)\theta_j}{\sin\theta_j}. \tag{3.5.14}$$

We now distinguish the cases $\left(\frac{p}{q}\right) = \pm 1$.

Case (i): $\left(\frac{p}{q}\right) = -1$. In this case we have seen that $X^{p,q}$ is bipartite and hence its eigenvalues are symmetric about zero. We have $\lambda_0 = p+1$, $\lambda_{n+1} = -(p+1)$ and $|\lambda_j| < p+1$ for $1 \le j \le n-2$ (since $X^{p,q}$ is connected). For k odd the right hand side of (3.5.13) is thus zero while for k even we have

$$\delta(p^k) + O_\epsilon(p^{k(1/2+\epsilon)}) = \frac{4}{n} \sum_{d|p^k} d + o(p^k) \qquad \text{as } k \to \infty.$$

Hence applying Lemma 3.5.1 we learn that

$$\delta(p^k) = \begin{cases} 0 & \text{if } k \text{ is odd} \\ \dfrac{4}{n} \displaystyle\sum_{d|p^k} d & \text{if } k \text{ is even.} \end{cases}$$

Eliminating this leading term in (3.5.13) we get

$$\frac{2\, p^{k/2}}{n} \sum_{j=1}^{n-2} \frac{\sin(k+1)\theta_j}{\sin\theta_j} = O_\epsilon(p^{k/2+\epsilon k}) \qquad \text{as } k \to \infty.$$

Hence

$$\sum_{j=1}^{n-2} \frac{\sin(k+1)\theta_j}{\sin\theta_j} = O_\epsilon(p^{k\,\epsilon}) \,.$$

The last clearly implies that θ_j, $1 \le j \le n-2$, are all real, i.e., $|\lambda_j| \le 2\sqrt{p}$ for $1 \le j \le n-2$. That is we have shown that for $\left(\frac{p}{q}\right) = -1$, $X^{p,q}$ is a bipartite Ramanujan graph.

Case (ii): $\left(\frac{p}{q}\right) = 1$. Now as we saw at the end of Section 3.3, $X^{p,q}$ is not bipartite. This time we have

$$\delta(p^k) + O_\epsilon(p^{k(1/2+\epsilon)}) = \frac{2\,(p^{k+1}-1)}{n\,(p-1)} + o(p^k) \,.$$

Hence by the Lemma

$$\delta(p^k) = \frac{2\,(p^{k+1}-1)}{n\,(p-1)}$$

and this time

$$\sum_{j=1}^{n-1} \frac{\sin(k+1)\theta_j}{\sin\theta_j} = O_\epsilon(p^{k\,\epsilon}) \quad \forall \epsilon > 0 \,.$$

Hence θ_j are real for $1 \le j \le n-1$, i.e., $X^{p,q}$ is a Ramanujan graph.

To prove $X^{p,q}$ is Ramanujan we used the full solution of the Ramanujan conjectures for weight two. Using the weaker bound proved in Proposition 1.5.3 would lead to $\lambda_1 \le p^{3/4} + p^{-3/4}$, which though not giving the complete truth is sufficient for many applications of these graphs.

Finally we can use the graphs $X^{p,q}$ to create a related family of Ramanujan graphs $Y^{p,q}$ as follows. $PGL(2, \mathbf{Z}/q\mathbf{Z})$ acts on $P^1(\mathbf{F}_q) = \{0, 1, \ldots, q-1, \infty\}$ in the usual linear fractional way. We turn $P^1(\mathbf{F}_q)$ into a $(p+1)$-regular graph by joining $\xi \in P^1$ to $\gamma\xi$ where $\gamma \in \{\alpha, \ldots, \overline{\alpha_s}\}$. This gives the graph $Y^{p,q}$. It has order $q+1$ and is regular of degree $p+1$ (it clearly will have loops!). Any eigenfunction f of A on $Y^{p,q}$ gives rise to one F on $X^{p,q}$ with the same eigenvalue. In fact $F(g) = f(g(0))$ supplies this correspondence. To show that $Y^{p,q}$ is Ramanujan (non-bipartite) we must show that $-(p+1)$ is not an eigenvalue of $Y^{p,q}$. If it were we clearly must be in the $\left(\frac{p}{q}\right) = -1$ case and we can assume $F(g) = f(g(0))$ is 1 on $PSL(2, \mathbf{Z}/q\mathbf{Z})$ and -1 on the complement. Now F is constant on the subgroup $\{\left(\begin{smallmatrix} \alpha & 0 \\ \gamma & \delta \end{smallmatrix}\right) \mid \alpha\delta \ne 0\}$. This subgroup clearly contains members of $PSL(2, \mathbf{Z}/q\mathbf{Z})$ as well as its complement which is a contradiction.

The graphs of $X^{2,3}, Y^{5,13}, Y^{5,17}$ are displayed in Figures 3.1 and 3.2.

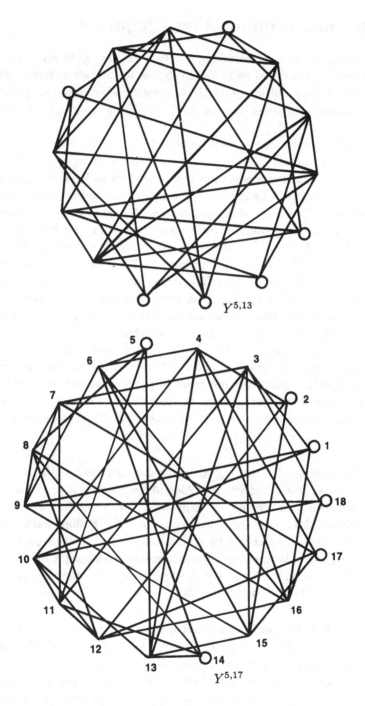

Figure 3.2(a) and Figure 3.2(b)

Notes and comments on Chapter 3

3.1. The problem of constructing graphs with large girth and large chromatic number was solved by Erdős [Er] by the method given in the text. For a history of the problem see [Bo1]. An explicit solution to this problem is due to Lubotzky–Phillips–Sarnak [LPS2,LPS3] via the graphs $X^{p,q}$ of this Chapter.

Expander graphs are the basic building blocks for the construction of superconcentrators and nonblocking networks, see Pippenger [Pi1]. The random construction of expanders was first carried out by Pinsker [Pin], see also Pippenger [Pi1] and Chung [Ch1]. The explicit expanders produced in this chapter are still not as good as what can be achieved by counting arguments. To see the difference consider the construction of expanders with, say, $3n$ edges. For each $0 < \alpha < 1$ one seeks graphs which expand sets of size αn in O to sets of size βn in I. Pippenger [Pi2] has obtained upper bounds for β as a function of α. The Ramanujan bipartite cubic graphs provide explicit graphs with expansion β as a function of α via Proposition 3.2.4. Also the random method provides a function β of α. Figure 3.3 is a plot of these functions $\beta(\alpha)$. α is on the x–axis and β on the y–axis. The lowest curve is the diagonal $y = x$. Next is the curve corresponding to the explicit bipartite Ramanujan graphs. The third curve is the random construction (Chung [Ch1]). Finally the minimum of the top two curves constitute an upper bound of the best possible expansion of sets of size αn, Pippenger [Pi2]. The plot was provided to us by Pippenger. There is clearly room for improving the explicit and random constructions.

3.2. Proposition 3.2.4 relating the expansion coefficient to the spectrum is due to Tanner [Tan]. The article of Alon [A1] explains in a very clear way the relation between eigenvalues and expanders. The bound for the diameter in Proposition 3.2.6 is essentially derived in Lubotzky–Phillips–Sarnak [LPS3]. For bounds on diameter and its applications as well as the construction of another family of interesting arithmetic graphs see Chung [Ch2]. The lower bound Proposition 3.2.7 is stated in Alon's paper [A1]; the proof here is due to Lubotzky–Phillips–Sarnak [LPS3].

3.3. and 3.4. The construction and analysis of the graphs $X^{p,q}$ carried out here follows closely the paper Lubotzky–Phillips–Sarnak [LPS3]. Margulis [Ma3,Ma4] has independently obtained a similar construction as well as some of the results in these Sections. In fact the first explicit construction of an expander graph (though without any determined expansion coefficient) is due to Margulis [Ma2] who made use of group representations and property T. For a discussion of the group representation approach as well as the

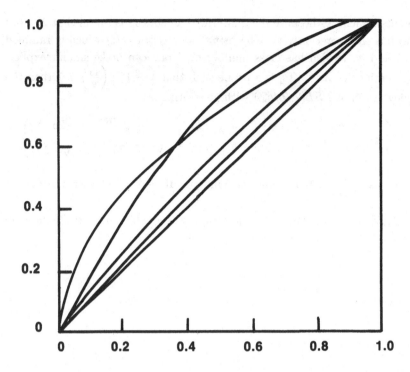

Figure 3.3: Pippenger's Plot

number theoretical ones see the article of Bien [Bi]. The recent article of de la Harpe and Valette [H–V] gives a nice discussion of property T and applications to expander graphs and invariant means.

Ihara [IH] constructs a family of Ramanujan graphs. The adjacency matrices of these graphs are essentially the 'Brandt matrices' [Ei2]. While these graphs may not be as explicit as the $X^{p,q}$'s they do carry very interesting number theoretic information. They are intimately connected to the reduction of the modular curves $X_0(\ell)$ (mod p) where ℓ is a prime unequal to p. Mestre [Me] has even used them very effectively to compute Fourier coefficients of holomorphic forms of weight 2 for $\Gamma_0(N)$ for large N.

Friedman [Fr] has recently shown that the random $2k$–regular graph (as $n \to \infty$) is close to being Ramanujan; he shows that

$$\lambda_1 \leq 2\sqrt{2k-1} + 2\log k + C.$$

Finally we note that one can easily modify the construction of the $X^{p,q}$ above to include the case of $p \equiv 3 \pmod 4$ (p prime). The prime $p = 2$

presents difficulties because the quaternion algebra H is ramified at 2. Chiu [Chi] has shown recently that by using the algebra D/Q which is ramified at 13 and ∞ (which has class number one) one can make similar explicit constructions of $X^{2,q}$. If q is a prime such that $\left(\frac{-2}{q}\right) = \left(\frac{13}{q}\right) = 1$ then the Cayley graph of $PSL(2, \mathbf{Z}/q\mathbf{Z})$ with generating set

$$S = \left\{ \begin{pmatrix} 1 & 0 \\ 0 & -1 \end{pmatrix}, \begin{pmatrix} 2-\sqrt{-2} & -\sqrt{-26} \\ -\sqrt{-26} & 2+\sqrt{-2} \end{pmatrix}, \begin{pmatrix} 2+\sqrt{-2} & \sqrt{-26} \\ \sqrt{-26} & 2-\sqrt{-2} \end{pmatrix} \right\},$$

in the case $\left(\frac{2}{q}\right) = 1$, is cubic Ramanujan. If $\left(\frac{2}{q}\right) = -1$ then the Cayley graph of
$PGL(2, \mathbf{Z}/q\mathbf{Z})$ with the same generating set is a cubic bipartite Ramanujan graph.

Chapter 4

Bounds for Fourier coefficients of 1/2–integral weight

In Proposition 1.5.5 we obtained an estimate for the Fourier coefficients of a cusp form of 1/2 integral weight. Precisely, the bound was

$$a(n) = O_\epsilon(n^{k/2-1/4+\epsilon}).$$

For applications, such as to the Linnik problem of Chapter 1, this bound falls just short of giving something non-trivial. Our aim in this chapter is to establish

Theorem 4.1. *Let $f(z)$ be a cusp form of half integral weight k for $\Gamma_0(N)$ (here $4 \mid N$) then for n square free we have*

$$|a_n| \ll_\epsilon n^{k/2-2/7+\epsilon}.$$

Remarks 4.2.

(1). The condition that n be square free or some related condition is necessary in view of the example in 1.3.4. In the notes and comments at the end of this chapter we explain how to extend the estimate of Theorem 4.1 to all n, as long as f is orthogonal to the theta functions of one variable (with respect to the Petersson inner product).

(2). Though Theorem 4.1 falls short of the Ramanujan conjecture 1.3.4, it does break the $k/2 - 1/4$ barrier. As such it allows one to solve the Linnik problem as follows:

From (1.3.11) it follows that if P is a homogeneous harmonic polynomial in $\mathbf{R}^3$ of degree $\nu \geq 1$, then

$$\theta_P(x) = \sum_{m \in \mathbf{Z}^3} P(m)\, e(|m|^2 z) = \sum_{n=1}^{\infty} a_n\, e(nz)$$

87

is a cusp form for $\Gamma_0(4)$ of weight $3/2 + \nu$. Hence by Theorem 4.1 we have for n square free

$$|a_n| \ll_\epsilon n^{3/4 + \nu/2 - 2/7 + \epsilon}. \tag{4.1}$$

Now

$$P(m/|m|) = |m|^{-\nu} P(m)$$

so that

$$a_n = n^{\nu/2} \sum_{|m|^2 = n} P\left(\frac{m}{|m|}\right)$$

hence

$$\sum_{|m|^2 = n} P\left(\frac{m}{|m|}\right) \ll_\epsilon (n^{13/28 + \epsilon}). \tag{4.2}$$

On the other hand by a theorem of Gauß[1] [Ga]

$$r_3(n) = \sum_{|m|^2 = n} 1 = \frac{24\, h(d)}{w(d)} \left(1 - \left(\frac{d}{2}\right)\right), \tag{4.3}$$

where $h(d)$ is the class number of $\mathbf{Q}(\sqrt{-n})$, $w(d) = \#$ of roots of 1 in this field, and $d = \mathrm{discr}\,(\mathbf{Q}(\sqrt{-n}))$. In this Linnik problem we are of course assuming $r_3(n) > 0$, i.e., $\left(\frac{d}{2}\right) = 1$. Now Siegel [Si2] has shown that $h(d) \gg_\epsilon |d|^{1/2 - \epsilon}$ (though non-effectively!). Hence for the n's under consideration we have

$$r_3(n) \gg_\epsilon n^{1/2 - \epsilon}. \tag{4.4}$$

It follows that for any P of degree $\nu \geq 1$

$$\frac{1}{r_3(n)} \sum_{|m|^2 = n} P(m/|m|) = O_\epsilon(n^{-1/28 + \epsilon}). \tag{4.5}$$

The harmonic polynomials on S^2 form an orthonormal basis and so it follows immediately that as $n \to \infty$

$$\frac{1}{r_3(n)} \sum_{|m|^2 = n} f(m/|m|) \longrightarrow \int_{S^2} f\, d\lambda \tag{4.6}$$

for all continuous functions f, that is to say the points $\{m|\, |m|^2 = n\}$ become equidistributed. The extension of (4.6) to the case of all n (for which $r_3(n) > 0$) is described in the notes at the end of this Chapter.

(3). Using the same theta function methods one can use Theorem 4.1 to deduce similar equidistribution results for other definite quadratic forms in three variables.

---

[1]This identity may also be derived from the Eisenstein series of weight $3/2$.

(4). For definite forms in more than three variables the equidistribution follows from the earlier $n^{k/2-1/4+\epsilon}$ bound. Indeed in these cases the singular series term is of order $n^{k/2-1}$.

We turn to the proof of Theorem 4.1. Let k be half an odd integer, $k \geq 5/2$ and let $4 \mid N$. Let $f_1, \ldots, f_R$ be an orthonormal basis of $S_k(\Gamma_0(N))$ (with respect to the Petersson inner product). From (1.5.3) we have

$$\langle P_m, f_j \rangle = \overline{a_j(m)} \frac{\Gamma(k-1)}{(4\pi m)^{k-1}}$$

where we have written

$$f_j(z) = \sum_{m=1}^{\infty} a_j(m) \, e(mz). \tag{4.7}$$

Also

$$P_m(z) = \sum_{j=1}^{R} \langle P_m, f_j \rangle f_j(z).$$

Hence

$$\hat{P}_m(n) = \sum_{j=1}^{R} \langle P_m, f_j \rangle a_j(n) = \frac{\Gamma(k-1)}{(4\pi m)^{k-1}} \sum_{j=1}^{R} \overline{a_j(m)} \, a_j(n). \tag{4.8}$$

We saw in (1.5.4) that $\hat{P}_m(n)$ may also be expressed in an infinite series involving Kloosterman sums. Combining (4.8) and (1.5.4) leads, on setting $n = m$, to the fundamental identity

$$\frac{\Gamma(k-1)}{(4\pi n)^{k-1}} \sum_{j=1}^{R_N} |a_j(n)|^2 = 1 + 2\pi i^{-k} \sum_{c \equiv 0 \,(\mathrm{mod}\, n)} \frac{K(n,n,c)}{c} J_{k-1}\left(\frac{4\pi n}{c}\right).$$
$$\tag{4.9}$$

The estimates of $|a_j(n)|$ relies on (4.9). Note the positivity of the left hand side which allows us to vary N when estimating $|a_j(n)|^2$. In order to exploit (4.9) we need to evaluate $K(n,n,c)$ in a form in which we can see cancellation.

Let $\kappa = 2k$, κ is odd. We have

$$K_k(m,n,c) = \sum_{d \,(\mathrm{mod}\, c)} \varepsilon_d^{-\kappa} \left(\frac{c}{d}\right) e\left(\frac{md + n\bar{d}}{c}\right).$$

Since ε_d depends on $d \pmod 4$, the sum is a little messy as far as the power of 2 dividing c. In fact a simple use of the Chinese remainder theorem and quadratic reciprocity yields

Lemma 4.3. *If $c = qr$ with $(q, r) = 1$ and $4 \mid r$ then*

$$K_k(m, n, c) = K_{k-q+1}(m\bar{q}, n\bar{q}, r)\, S(m\bar{r}, n\bar{r}, q)\,,$$

where for q odd $S(m, n, q)$ is the Salié sum

$$S(m, n, q) = \sum_{x \ (\mathrm{mod}\ q)} \left(\frac{x}{q}\right) e\left(\frac{mx + n\bar{x}}{q}\right)\,. \qquad (4.10)$$

Thus after removing the messy power of 2 part, the K_k is essentially a Salié sum. The point here is unlike the seemingly simpler Kloosterman sum $K(m, n, c)$ the Salié sum $S(m, n, q)$ may be evaluated in elementary terms. The situation is the finite analog of the Bessel function $J_{k-1}(z)$ being an elementary function when k is half an odd integer; for example

$$J_{1/2}(z) = \sqrt{\frac{2}{\pi z}}\, \sin z\,. \qquad (4.10a)$$

Lemma 4.4. *Let $(m, q) = (n, q) = 1$*

(i) $$S(m, n, q) = \left(\frac{m}{n}\right) S(1, mn, q)$$

(ii) $$S(1, m, q) = 0 \qquad \text{if } \left(\frac{m}{q}\right) = -1$$

(iii) $$S(1, n^2, q) = \varepsilon_q \sqrt{q} \sum_{x^2 \equiv 1 \ (\mathrm{mod}\ q)} e\left(\frac{2xn}{q}\right)$$

Corollary 4.5. *q odd and $(n, q) = 1$ then*

$$S(n, n, q) = \left(\frac{n}{q}\right) \varepsilon_q \sqrt{q} \sum_{\substack{ab=q \\ (a,b)=1}} e\left(2n\left(\frac{\bar{a}}{b} - \frac{\bar{b}}{a}\right)\right)\,.$$

Proof of Corollary 4.5: From Lemma 4.4 we have

$$S(n, n, q) = \left(\frac{n}{q}\right) S(1, n^2, q) = \left(\frac{n}{q}\right) \varepsilon_q \sqrt{q} \sum_{x^2 \equiv 1 \ (\mathrm{mod}\ q)} e\left(\frac{2xn}{q}\right)\,.$$

It is straightforward that

$$\sum_{x^2 \equiv 1 \ (\mathrm{mod}\ q)} e\left(\frac{2xn}{q}\right) = \sum_{\substack{ab=q \\ (a,b)=1}} e\left(2n\left(\frac{\bar{a}}{b} - \frac{\bar{b}}{a}\right)\right)\,. \qquad \square$$

Proof of Lemma 4.4:

(i)
$$S(m,n,q) = \sum_{x \,(\text{mod}\, q)} \left(\frac{x}{q}\right) e\left(\frac{mx + n\bar{x}}{q}\right).$$

Let $y = mx$, $x = \bar{m}y$, $\bar{x} = m\bar{y}$, then

$$S(m,n,q) = \sum_{x \,(\text{mod}\, q)} \left(\frac{\bar{m}y}{q}\right) e\left(\frac{y + nm\bar{y}}{q}\right) = \left(\frac{m}{q}\right) S(1, mn, q).$$

(ii)
$$S(1,m,q) = \sum_{x \,(\text{mod}\, q)} \left(\frac{x}{q}\right) e\left(\frac{x + m\bar{x}}{q}\right)$$

$$= \sum_{x \,(\text{mod}\, q)} \left(\frac{\bar{x}}{q}\right) e\left(\frac{\bar{x} + mx}{q}\right)$$

$$= S(m, 1, q)$$

$$= S(1, m, q) \left(\frac{m}{q}\right) \qquad \text{by } (i).$$

Hence $S(1, m, q) = 0$ if $\left(\frac{m}{q}\right) = -1$.

(iii) Now take $m = n^2$, let

$$h(n) = \sum_{x \,(\text{mod}\, q)} \left(\frac{x}{q}\right) e\left(\frac{x + n^2\bar{x}}{q}\right)$$

for $n = 0, 1, \ldots, q-1$. h is a function on the group $\mathbf{Z}/q\mathbf{Z}$ and we compute its Fourier transform

$$\hat{h}(m) = \sum_{n \,(\text{mod}\, q)} h(n) e\left(\frac{-mn}{q}\right)$$

$$= \sum_{x} \left(\frac{x}{q}\right) e\left(\frac{x}{q}\right) \sum_{n} e\left(\frac{\bar{x}\,(n^2 - xmn)}{q}\right)$$

$$= \sum_{x} \left(\frac{x}{q}\right) e\left(\frac{x}{q}\right) \sum_{n} e\left(\frac{\bar{x}\,(n - xm/2)^2 - (xm/2)^2\,\bar{x}}{q}\right)$$

$$= \sum_{x} \left(\frac{x}{q}\right) e\left(\frac{x}{q}\right) \sum_{n} e\left(\frac{-xm^2)}{4}\right) \varepsilon_q \sqrt{q} \left(\frac{\bar{x}}{q}\right)$$

by the standard evaluation of the Gauß sum. Hence

$$\hat{h}(m) = \varepsilon_q \sqrt{q} \sum_{(x,q)=1} e\left(\frac{x}{q}\left(1 - \frac{m^2}{4}\right)\right). \tag{4.11}$$

The sums

$$c_q(r) = \sum_{\substack{(x,q)=1 \\ x \,(\mathrm{mod}\,q)}} e\left(\frac{xr}{q}\right) \tag{4.12}$$

were encountered in (1.4.8) and are known as Ramanujan sums. As Ramanujan observed these may be evaluated by Möbius inversion [R2]

$$c_q(r) = \sum_{d|q,\ d|r} \mu(n/d)\, d, \tag{4.13}$$

where μ is the Möbius function.

Hence using Fourier inversion we have

$$
\begin{aligned}
h(n) &= \frac{1}{q} \sum_{m \,(\mathrm{mod}\,q)} \hat{h}(m) e\left(\frac{mn}{q}\right) \\
&= \frac{\varepsilon_q}{\sqrt{q}} \sum_{m \,(\mathrm{mod}\,q)} \sum_{d|q,\ d|(m^2-4)} d\,\mu(q/d)\, e\left(\frac{mn}{q}\right) \\
&= \frac{\varepsilon_q}{\sqrt{q}} \sum_{d|q} d\,\mu(q/d) \sum_{\substack{m \,(\mathrm{mod}\,q) \\ m^2 \equiv 4 \,(\mathrm{mod}\,d)}} e\left(\frac{mn}{q}\right).
\end{aligned}
\tag{4.14}
$$

Now assuming, as we are, that $(n,q) = 1$ we claim that

the inner sum in (4.14) is zero for every $d\,|\,q,\ d \neq q$. $\qquad$ (4.15)

With (4.15) established it follows that

$$h(n) = \sqrt{q}\,\varepsilon_q \sum_{m^2 \equiv 4 \,(\mathrm{mod}\,q)} e\left(\frac{mn}{q}\right)$$

which proves part (iii) of Lemma 4.4.

To see (4.15) let $q = db$ with $b > 1$. For each $m_0\ (\mathrm{mod}\,d)$ a solution of $m^2 \equiv 4\ (\mathrm{mod}\,d)$ we have solutions m of $m^2 \equiv 4\ (\mathrm{mod}\,d)$, $m\ (\mathrm{mod}\,q)$, where $m = m_0 + \lambda d$, $\lambda\ (\mathrm{mod}\,b)$. Hence the inner sum breaks up into sums of the type

$$\sum_{\lambda \,(\mathrm{mod}\,b)} e\left(\frac{(m_0 + \lambda d)n}{db}\right) = e\left(\frac{m_0 n}{db}\right) \sum_{\lambda \,(\mathrm{mod}\,b)} e\left(\frac{\lambda n}{b}\right).$$

Since $(n,b) = 1$ this last sum is zero as claimed. $\qquad\square$

With this evaluation of $K(n,n,c)$ from Corollary 4.5 and Lemma 4.3 we can turn to estimating the right hand side of 4.9, in n. The variation of $J_{k-1}(4\pi n/c)$ is moderate and well understood and not really the issue here.

Rather it is the variation in the sign of $K(n, n, c)$ that is arithmetic and must be exploited. Since by Lemma 4.3 $K(m, n, c)$ is essentially $S(n, n, q)$ the issue is really one of estimating sums of the model problem:

$$K_N(n, x) = \sum_{\substack{q \leq x \\ q \equiv 0 (\mathrm{mod}\, N)}} \frac{S(n, n, q)}{\sqrt{q}}. \tag{4.16}$$

Model Problem.

We want to develop non-trivial estimates for $K_N(n, x)$ above by exploiting cancellation

$$K_N(x) = \sum_{\substack{q \leq x \\ q \equiv 0\, (\mathrm{mod}\, N)}} \left(\frac{n}{q}\right) \sum_{\substack{ab=q \\ (a,b)=1}} e\left(2n\left(\frac{\bar{a}}{b} - \frac{\bar{b}}{a}\right)\right). \tag{4.17}$$

In what follows we make estimates for $K_N(x)$ uniformly in the variables x, N, and n. We assume n is large and square free and is the primary variable, x, N will be chosen to depend on n (eventually) but for the time being we consider them to be $O(n^r)$ for some fixed large r. Any quantity which is $O_\epsilon(x^\epsilon n^\epsilon N^\epsilon)$ for all $\epsilon > 0$ will be denoted $\mathcal{L}$. Thus for example $\tau(n) = \#$ divisors of $n = O(\mathcal{L})$. The trivial bound for $K_N(x)$ is clearly

$$|K_N(x)| \leq \frac{x}{N}\mathcal{L} + \mathcal{L}. \tag{4.18}$$

The idea in improving over (4.18) is as follows. In the range where either a or b is small there is not much gained in cancellation from the exponential terms in (4.17) but in this case the $\left(\frac{n}{q}\right)$ term will give cancellation since we are assuming n is square free! When a and b are both large we will exploit the cancellation coming from the exponential terms in (4.17). In order to divide the sum into such ranges we write

$$K_N(y) = \sum_{\substack{y < q < 2y \\ q \equiv 0\, (\mathrm{mod}\, N)}} \left(\frac{n}{q}\right) \sum_{ab=q} e\left(2n\left(\frac{\bar{a}}{b} - \frac{\bar{b}}{a}\right)\right). \tag{4.19}$$

There are at most $\log x = O(\mathcal{L})$ such series so we need deal only with one at a time.

Now

$$K_N(y) = \sum_{\substack{y < ab < 2y \\ (a,b)=1 \\ N|ab}} \left(\frac{n}{ab}\right) e\left(2n\left(\frac{\bar{a}}{b} - \frac{\bar{b}}{a}\right)\right). \tag{4.20}$$

So again dividing into a logarithmic number of series we consider the basic sum that we need to estimate

$$F(A, B, N) = \sum_{\substack{y < ab < 2y \\ (a,b)=1 \\ N|ab \\ A \leq a \leq 2A \\ B \leq b \leq 2B}} \left(\frac{n}{ab}\right) e\left(2n\left(\frac{\bar{a}}{b} - \frac{\bar{b}}{a}\right)\right). \qquad (4.21)$$

Incomplete sum.

The dependence of $e\left(2n\left(\frac{\bar{a}}{b} - \frac{\bar{b}}{a}\right)\right)$ on a, or on b, can be made more transparent by the following very useful reciprocity observed by Iwaniec:
Since clearly $a\bar{a} + b\bar{b} \equiv 1 \pmod{ab}$, we have

$$\frac{\bar{a}}{b} + \frac{\bar{b}}{a} \equiv \frac{1}{ab} \pmod{1}. \qquad (4.22)$$

For ab large we can then use this to approximately flip a and b, thus

$$e\left(2n\left(\frac{\bar{a}}{b} - \frac{\bar{b}}{a}\right)\right) = e\left(\frac{4n\bar{a}}{b}\right)\left(1 + O\left(\frac{n}{y}\right)\right). \qquad (4.23)$$

Now for sums of the type

$$\sum_{A \leq a \leq 2A} \left(\frac{n}{ba}\right) e\left(4n\left(\frac{\bar{a}}{b}\right)\right), \qquad (4.23')$$

we can make non-trivial bounds by completing the sum and using bounds from algebraic geometry. Precisely, let $F(u)$ be a function on the integers which is periodic of period m. Then

$$\sum_{u \leq X} F(u) = \sum_{u \leq X} \frac{1}{m} \sum_{r \pmod m} \hat{F}(r) e\left(\frac{ru}{m}\right)$$

$$= \frac{X}{m} \hat{F}(0) + \sum_{\substack{r \pmod m \\ r \neq 0}} \frac{1}{m} \hat{F}(r) \sum_{u \leq X} e\left(\frac{ru}{m}\right)$$

$$\ll \frac{X|\hat{F}(0)|}{m} + \frac{1}{m} \sum_{\substack{r \pmod m \\ r \neq 0}} \frac{|\hat{F}(r)|}{|1 - e(r/m)|}$$

$$\ll \frac{X|\hat{F}(0)|}{m} + \frac{\|\hat{F}\|_\infty}{m} \sum_{1 \leq r \leq m/2} \frac{1}{r}$$

i.e.,

$$\sum_{u \leq X} F(u) \ll \frac{X \, |\hat{F}(0)|}{m} + \|\hat{F}\|_\infty \log m \,. \qquad (4.24)$$

The above method is known as completing the sum. The term $\frac{X \, \hat{F}(0)}{m}$ clearly corresponds to the sum over a period while we gain in the second term if $\|\hat{F}\|_\infty$ is small.

To apply this consider for example

$$\sum_{u \leq X} e\left(\frac{\nu \bar{u}}{m}\right) \qquad (4.25)$$

$$\hat{F}(r) = \sum_{u \, (\mathrm{mod}\, m)} e\left(\frac{\nu \bar{u} + ru}{m}\right) \qquad (4.25)$$

which is a Kloosterman sum. From the Weil estimate in Chapter 1

$$|\hat{F}(r)| \leq (\nu, m)^{1/2} m^{1/2} \tau(m) \,. \qquad (4.27)$$

Hence

$$\sum_{u \leq X} e\left(\frac{\nu \bar{u}}{m}\right) \ll \frac{\tau(m) \, X(\nu, m)}{m} + (m, \nu)^{1/2} m^{1/2} \tau(m) \log m \,. \qquad (4.28)$$

Hence the sum in (4.23′)

$$\sum_{a \leq A} \left(\frac{n}{ba}\right) e\left(\frac{4n \bar{a}}{b}\right) \ll (bn)^{1/2} \mathcal{L} \qquad (4.29)$$

where we have applied (4.24) and used the Weil bound for the complete sum of period bn and also the important point that if n is square free

$$\sum_{a \, (\mathrm{mod}\, bn)} \left(\frac{n}{ba}\right) e\left(\frac{4n \bar{a}}{b}\right) = 0 \,.$$

With these remarks we turn to estimating $F(A, B, N)$ in (4.21).

When A or B is small.

If say B is small we carry out the a sum in (4.21) first and use the method in the previous Section. Using (4.23) and summation by parts we find for each fixed b having to estimate

$$\sum_{A_1 \leq a \leq A_2} \left(\frac{n}{a}\right) e\left(4n \frac{\bar{a} \, m}{b}\right)$$

A_1 and A_2 are the range of $a' = a/(N, b)$ and m depends on b. From (4.29) we thus find that

$$|F(A, B, N)| \ll B\, B^{1/2} n^{1/2} \mathcal{L} \left(1 + \frac{n}{y}\right)$$

$$= B^{3/2} n^{1/2} \mathcal{L} \left(1 + \frac{n}{y}\right) \tag{4.30}$$

On the other hand doing the sum on b first gives

$$|F(A, B, N)| \ll A^{3/2} n^{1/2} \mathcal{L} \left(1 + \frac{n}{y}\right). \tag{4.31}$$

The bounds (4.30) and (4.31) are good only if either A or B is small. The factor $n^{1/2}$ appears because of the need to exploit the sign changes in $\left(\frac{n}{a}\right)$ which makes the period of the complete sum, of length nb. When both A and B are large we forego this cancellation and apply Cauchy–Schwarz. However a direct attempt does not suffice to give the result. We will overcome this by one more trick of embedding.

When A and B are large

It is not sufficient to bound $F(A, B, N)$ for fixed N in this case. However if we average, say

$$F_P = \sum_{P \leq p \leq 2P} |F(A, B, p)|, \tag{4.32}$$

where p runs over primes in the range $P, 2P$ we can sum as follows:

$$\sum_{P \leq p \leq 2P} \sum_{\substack{A \leq ab \leq 2A \\ B \leq b \leq 2B \\ (a,b)=1 \\ y \leq ab \leq 2y \\ p|ab}} \lambda_p \left(\frac{n}{ab}\right) e\left(2n\left(\frac{\overline{a}}{b} - \frac{\overline{b}}{a}\right)\right)$$

where $\lambda_p = \operatorname{sgn} F(A, B, p)$.

Now since $p|a$ or $p|b$ we have

$$F_P(A, B) = F(A/p, B) + F(A, B/p)$$

and we may estimate these separately. Changing variables

$$|F(A/p, B)| \leq \sum_{A/p \leq a \leq 2A/p} \sum_{B \leq b \leq 2B} \left| \sum_{P_1 \leq p \leq P_2} \lambda_p \left(\frac{n}{p}\right) e\left(2n\left(\frac{\overline{ap}}{b} - \frac{\overline{b}}{ap}\right)\right) \right|$$

where $[P_1, P_2]$ is the appropriate range (its length $\ll P$). So by Cauchy–Schwarz

$$|F(A/p, B)|^2 \leq \frac{AB}{P^{-1}} \sum_{P_1, P_2} \left| \sum_{a,b} e\left(2n(p_2 - p_1)\left(\frac{\overline{ap_1 p_2}}{b} - \frac{\overline{b}}{ap_1 p_2}\right)\right)\right|.$$

(4.33)

Again if say we do the a sum first and use the reciprocity (4.22) we get for $p_1 \neq p_2$

$$\sum_a \leq \left(1 + \frac{n}{y}\right) \mathcal{L}\left((p_2 - p_1, b)^{1/2} + \frac{(p_2 - p_1, b)\, A}{bp}\right).$$

(4.34)

The sum in (4.33) for $p_1 = p_2$ gives

$$\frac{AB}{P} P \frac{A}{P} B = \frac{(AB)^2}{P}$$

while from (4.34) and

$$\sum_{B \leq b \leq 2B} \frac{(p_2 - p_1, b)}{b} \ll \mathcal{L}$$

we have the sum in (4.33) with $p_1 \neq p_2$ is

$$\ll \frac{AB}{P} P^2 \left(1 + \frac{n}{y}\right) \mathcal{L}\left(B^{3/2} + \frac{A}{P}\right).$$

(4.35)

Combining (4.34) and (4.35) gives

$$|F(A/P, B)| \ll \frac{AB}{P^{1/2}} + \left(1 + \frac{n}{y}\right)^{1/2} \mathcal{L}\, (A^{1/2}B^{5/4}P^{1/2} + AB^{1/2}).$$

(4.36)

On the other hand doing the b sum first gives

$$|F(A/P, B)| \ll \frac{AB}{P^{1/2}} + \left(1 + \frac{n}{y}\right)(A^{5/4}B^{1/2}P^{1/4} + A^{1/2}B\,P^{-1/2})\mathcal{L}.$$

(4.37)

Combining these gives (and doing the same for $F(A, B/P)$)

$$F_P(A, B) \ll \frac{y}{P^{1/2}} + \left(1 + \frac{n}{y}\right)^{1/2}(y^{7/8}P^{3/8} + (A^{-1/2} + B^{-1/2})y)\mathcal{L}.$$

(4.37)

We note that in order to be effective here we must choose P large (say a small power of n) and A and B large in view of the $(A^{-1/2} + B^{-1/2})$ term. Notice that in (4.37) we do not have the bad power of $n^{1/2}$ since we do not use the sign of $\left(\frac{n}{\cdot}\right)$; we are exploiting the exponential part only!

In fact optimizing in (4.37), (4.30), and (4.31) we divide between the cases

$$A \text{ or } B \leq \left(1 + \frac{n}{y}\right)^{-1/4} n^{-1/4} y^{1/2} P^{-1/2}$$

when we apply (4.30) or (4.31), else we use (4.37). This gives

$$\sum_{P \leq p \leq 2P} |F(A, B, p)| \ll \frac{y}{P^{1/2}} + \left(1 + \frac{n}{y}\right)^{5/8} (y^{7/8} P^{3/8} + n^{1/8} y^{3/4} P^{1/4}) \mathcal{L}.$$

Returning to (4.17) we get

$$\sum_{P \leq p \leq 2P} |K_p(x)|$$
$$\ll \left[\frac{x}{P^{1/2}} + \frac{x}{n^{1/2}} + (x + n)^{5/8} (x^{1/4} P^{3/8} + n^{1/8} x^{1/8} P^{1/4})\right] \mathcal{L}. \quad (4.38)$$

This is the basic bound for the model problem. To deal with the actual sums involving $K_k(m, n, c)$ viz.

$$\hat{K}_Q(x) = \sum_{\substack{c \leq x \\ c \equiv 0 \,(\mathrm{mod}\, Q)}} c^{-1/2} K(n, n, c) e\left(\frac{2\nu n}{c}\right), \quad (4.39)$$

where $\nu = -1, 0, 1$, one needs to modify the above taking Lemma 4.3 into account (i.e., factoring out the power of 2 in Lemma 4.3), see Iwaniec [Iw1]. The result one gets in this way is

Proposition 4.6. *For n square free and N_0 fixed*

$$\sum_{P \leq p \leq 2P} |\hat{K}_{N_0 p}(x)|$$
$$\ll \mathcal{L} [x P^{-1/2} + x n^{-1/2} + (x + n)^{5/8} (x^{1/4} P^{3/8} + n^{1/8} x^{1/8} P^{1/4})].$$

Proof of Theorem 4.1: The reason the above averaging over the level in the last estimate can be exploited is the simple but beautiful observation that if $f(z)$ is our given orthonormal cusp form for $\Gamma_0(N)$ then $(1/[\Gamma_0(N), \Gamma_0(pN)]^{1/2}) f(z)$ is also such a form for $\Gamma_0(pN)$. Hence for each p we can arrange to have the term $|a(n)|^2/[\Gamma_0(N), \Gamma_0(pN)]$ appear on the left of (4.9), where $a(n)$ is the coefficient to be estimated. Now $[\Gamma_0(N), \Gamma_0(pN)] \leq p + 1$ so summing $P \leq p \leq 2P$ gives

$$(\log P)^{-1} \frac{|a(n)|^2}{n^{k-1}} \ll \frac{P}{\log P} + \sum_{P \leq p \leq 2P} \left| \sum_{c \equiv 0 \,(\mathrm{mod}\, p N_0)} \frac{K(m, n, c)}{c} J_{k-1}\left(\frac{4\pi n}{c}\right) \right|.$$

Using the evaluation as in (4.10′) but for the more general case [Wa],

$$J_{k-1}(2\pi z) = \left(\frac{2}{z}\right)^{1/2}\left(e(z)\,H_1\left(\frac{2}{z}\right) + e(-z)\,H_{-1}\left(\frac{2}{z}\right)\right)$$

where H_1 and H_{-1} are polynomials of degree $\leq k - 1/2$, for $z \geq 2$ and simply $J_{k-1}(z) = O(z^{k-1})$ and $J'_{k-1}(z) = O(z^{k-2})$ for $0 < z < 2$ we can express the series

$$\sum_{c \equiv 0 \,(\mathrm{mod}\, p\,N_0)} \frac{K(n,n,c)}{c}\, J_{k-1}\left(\frac{4\pi n}{c}\right) = \sum_{c \leq n} + \sum_{c > n}$$

as integrals against $\hat{K}_{p\,N_0}(x)$. Plugging in the basic estimate (after integration by parts) in Proposition 4.6 with $P = n^{1/7}$ then leads directly to

$$|a(n)|^2 \ll_\epsilon n^{k-1+3/7+\epsilon}$$

(see Iwaniec [Iw1]) which proves the Theorem.   □

Notice that the case A or B small exploits the sign of $\left(\frac{n}{ab}\right)$ since the solutions to $x^2 \equiv 1 \pmod{q}$ are not equidistributed, but for A and B large we exploit the equidistribution of the solutions to $x^2 \equiv 1 \pmod{q}$.

Notes and comments on Chapter 4

(a). The first remark we wish to make concerns the extension of Theorem 4.1 and also the cases of the Linnik conjecture to the general integer n. To do so we assume that the reader is familiar with the 'Shimura lift' described in Shimura's paper [Sh]. (Actually there are some difficulties in that paper in defining the lift because of the Hecke operators T_{p^2} with p dividing the level of the form but these may be easily taken care of by the theory of New Forms [AL].) If $f \in S_k(\Gamma_0(N))$, k half integral, and if f is an eigenform of the Hecke operators T_{p^2} then Shimura shows that for t square free (as usual $f = \sum a_n\, e(nz)$)

$$\sum_{n=1}^{\infty} a(t\,n^2)\, n^{-s}$$

$$= a(t) \prod_p \left(1 - \left(\frac{t}{p}\right)\chi_1(p)\, p^{k-3/2-s}\right)(1 - w_p\, p^{-s} + p^{2k-2-2s})^{-1}$$

where $\chi_1(m) = \left(\frac{-1}{m}\right)^{k-1/2}$, and moreover if

$$\sum_{n=1}^{\infty} \frac{b_n}{n^s} = \prod_p (1 - w_p\, p^{-s} + p^{2k-2-2s})^{-1}$$

then

$$\sum_{n=1}^{\infty} b_n \, e(nz) \triangleq F(z) \in S_{2k-1}(\Gamma_0(N'))$$

for suitable N' (as long as f is orthogonal to the theta functions of one variable [Sh]). In particular if $k \geq 5/2$ this is so.

It follows that

$$a(t \, n^2) = a(t) \sum_{d|n} \chi(d) \, \mu(d) \, d^{k-3/2} b\left(\frac{n}{d}\right) .$$

The coefficients $b(n)$ are those of a modular form of even integral weight $2k - 1$ and by the Ramanujan conjectures for these

$$b(n) \ll_\epsilon n^{k+\epsilon} .$$

Hence

$$|a(t \, n^2)| \ll_\epsilon |a(t)| \sum_{d|n} d^{k-3/2} \left|\frac{n}{d}\right|^{k-1+\epsilon}$$

$$= |a(t)| \, n^{k-1+\epsilon} \sum_{d|n} d^{-1/2}$$

$$\ll_\epsilon |a(t)| \, n^{k-1+\epsilon} .$$

Hence applying Theorem 4.1

$$|a(t \, n^2)| \ll_\epsilon (t \, n^2)^{k/2 - 2/7 + \epsilon} . \tag{4.N.1}$$

This gives the extension of Theorem 4.1 to the general integer n.

(b). There is an analog of the Linnik problem for indefinite forms in three variables. This amounts to investigating the distribution of binary quadratic forms. Precisely let $d < 0$ be a discriminant of a binary quadratic form $[a, b, c] = a \, x^2 + bxy + cy^2$. As was shown by Gauß [Ga] these forms split under the action of $SL(2, \mathbf{Z})$ (by linear change of variable on (x, y)) into a finite number $h(d)$ (the class number) of inequivalent forms. We can associate with each form $[a, b, c]$ above a point $z \in \mathbf{H}$ by taking z as the solution of $a \, z^2 + b \, z + c = 0$. Then a set of representatives for the $h(d)$ reduced forms can be chosen to lie in the fundamental domain $\mathcal{F}$ in Figure 1.1. As $d \to \infty$, the question is that of the distribution of these points in $\Gamma(1)\backslash\mathbf{H}$. Linnik was able to show conditionally that these become equidistributed with respect to the invariant measure on $\Gamma(1)\backslash\mathbf{H}$ viz. $dxdy/y^2$. In a beautiful paper Duke [Du] has proved the result unconditionally. To do so he develops estimates similar to those of Theorem 4.1 but for the Fourier coefficients of the general half integral weight Maaß form as defined in A.2.4.

The estimations in this chapter exploit the analysis of the solutions to $x^2 \equiv 1 \pmod{c}$ with varying c. The analogous question for $x^2 \equiv m \pmod{c}$, where m is not a perfect square has a different behavior but one which is also closely linked to modular forms. The method of Appendix 2.1 of Chapter 1 applies equally well here and yields

$$\sum_{c \leq X} \frac{K_k(m, n, c)}{c} = O_\epsilon(X^{1/6+\epsilon}).$$

Now as we have seen from Lemma 4.4 and 4.3 these sums $K_k(m, n, c)$ are closely related to the sums

$$\sum_{x^2 \equiv m \pmod{c}} e\left(\frac{2nx}{c}\right),$$

that is to say 'Weyl sums' for the equidistribution of $x^2 \equiv m \pmod{c}$ with varying c. This should be compared with Hooley [Hoo] and Hejhal [He]. The more interesting question of the distribution of the roots of $x^2 \equiv -1 \pmod{p}$ for $p \leq X$ a prime, seems out of the reach of these methods. However very little is known. For example besides the limit point of zero for x/p where $x^2 \equiv -1 \pmod{p}$, $p \to \infty$ we know of no other limit point in $[0, 1/2)$.

Bibliography

[A1]   N. Alon, *Eigenvalues and expanders*, Combinatorica **6** (1986), 83–96.

[AM]   N. Alon and V.D. Millman, λ_1 *isoperimetric inequalities for graphs and superconcentrators*, Journal Comb. Theory, Ser. B **38** (1985), 73–88.

[AL]   A. Atkin and J. Lehner, *Hecke operators on* $\Gamma_0(m)$, Math. Ann. **185** (1970), 134–160.

[Ba]   S. Banach, *Sur le problème de la mesure*, Fund. Math. **4** (1923), 7–33.

[Bi]   F. Bien, *Construction of telephone networks by group representations*, Notices of the A.M.S. **36** (1989), 5-22.

[Bo1]   B. Bollobas, *Extremal graph theory*, Academic press, London 1978.

[Bo2]   B. Bollobas, *Random Graphs*, Academic Press, London 1985.

[BFH]   D. Bump, S. Friedberg, and J. Hoffstein, *Eisenstein series on the Metaplectic group and non vanishing theorems for automorphic L-functions and their derivatives*, Annals of Math. **131** (1990), 53-128.

[Chi]   P. Chiu, *The cubic Ramanujan graph*, Preprint 1989.

[Ch1]   F. Chung, *On concentrators, superconcentrators, generalizers, and non blocking networks*, Bell Sys. Tech. J. **59** (1978), 1765–1777.

[Ch2]   F. Chung, *Diameters and eigenvalues*, Journal of A.M.S. **2** (1989), 187–200.

[Da1]   H. Davenport, *Multiplicative Number Theory*, Second Edition, revised by H.L. Montgomery, Springer–Verlag, New York 1980.

[Da2]   H. Davenport, *Analytic methods for Diophantine equations and inequalities*, Ann Arbor Publ. 1962.

[De]   P. Deligne, *La conjecture de Weil I*, Publ. I.H.E.S. **43** (1974), 273–307.

[DS]   P. Deligne and J.P. Serre, *Formes modulaires de poids 1*, Ann. Sci. Ens. **7** (1974), 507–530.

[DI]   J.M. Deshouillers and H. Iwaniec, *Kloosterman sums and Fourier coefficients of cusp forms*, Inv. Math. **70** (1982), 219–288.

[Di]   L.E. Dickson, *Arithmetic of quaternions*, Proc. London Math. Soc. (2) **20** (1922), 225–232.

[Dr]   V.G. Drinfeld, *Finitely additive measures on S^2 and S^3 invariant with respect to rotations*, Func. Anal. Appl. 18, No 2 (1984), p. 77.

[Du]   W. Duke, *Hyperbolic distribution problems and half–integral weight Maaß forms*, Inv. Math. **92** (1988), 73–90.

[Ei1]   M. Eichler, *Quaternäre quadratische Formen und die Riemannsche Vermutung für die Kongruentzzeta Funktion*, Archiv. der Math. **5** (1954), 355–366.

[Ei2]   M. Eichler, *The basis problem for modular forms and the traces of the Hecke operators*, Springer Lecture Notes **320**, (1973), 77–151.

[EGM]   J. Elströdt, F. Grünewald, and J. Mennicke, *Poincaré series, Kloosterman sums and eigenvalues of the Laplacian for congruence groups acting on hyperbolic space*, C. R. Acad. Sci. Paris, t.305, I (1987), 577–581.

[Er]   P. Erdős, *Graph theory and probability*, Can. J. Math. **11** (1959), 34–38.

[Fo]   G. Folland, *Introduction to partial differential equations*, Princeton University Press 1976.

[Fr]   J. Friedman, *On the second eigenvalue and random walks in random d–regular graphs*, Preprint 1988.

[Ga]   C.F. Gauß, *Disquisitiones arithmeticae*, Leipzig, Fleisher (1801), English translation, Springer-Verlag.

[GJ]   S. Gelbart and H. Jacquet, *A relation between automorphic representations of $GL(2)$ and $GL(3)$*, Ann. Sci. Ecole Norm. Sup. **11** (1978), 471–542.

[GS]   D. Goldfeld and P. Sarnak, *Sums of Kloosterman sums*, Inv. Math. **71** (1983), 243–250.

[GR]   I. Gradsteyn and I.M. Ryzhik, *Tables of integrals, series, and products*, Academic press, New York 1980.

[Gr]   E. Granierer, *Criteria for the compactness and for discreteness of locally compact amenable groups*, Proc. A.M.S. **40** (1973), No. 2.

[H]   E. Hecke, *Mathematische Werke*, Göttingen 1959.

[Hoo]   C. Hooley, *On the distribution of roots of polynomial congruences*, Mathematika **11** (1964), 39–49.

[H1]   R. Howe, *θ–series and invariant theory*, Proc. Symp. A.M.S. XXXIII (1979), 275–286.

[HP]   R. Howe and I. Piatetski–Shapiro, *A counter example to the generalized Ramanujan conjecture for quasi split groups*, Proc. Symp. A.M.S. XXXIII (1979), 315–322.

[He]   D. Hejhal, *Roots of quadratic congruences and eigenvalues of the non Euclidian Laplacian*, Cont. Math. **53** (1985), 277–339.

[H-E]   P. de la Harpe and A. Valette, *La propriété (T) de Kazhdan pour les groupes localement compacts*, Astérisque **175** (1989).

[Ig]   J. Igusa, *Fibre systems of Jacobian varieties III*, Amer. Journal **81** (1959), 453–476.

[Ih]   Y. Ihara, *Discrete subgroups of $PGL(2, k_p)$*, Proc. Symp. in Pure Math. IX, A.M.S. 1968, 272–278.

[Iw1]   H. Iwaniec, *Spectral theory of automorphic functions and recent developments in analytic number theory*, I.C.M. Berkeley Proceedings (1986), 444–456.

[Iw2]   H. Iwaniec, *Fourier coefficients of modular forms of half integral weight*, Inv. Math. **87** (1987), 385–401.

[Iw3]   H. Iwaniec, *Small eigenvalues for congruence groups*, to appear in Acta Arithmetica.

[JL]   H. Jacquet and R. Langlands, *Automorphic forms on $GL(2)$*, Springer Lecture Notes **114** (1970).

[Ka]   D. Kazhdan, *Connection of the dual space of a group with the structure of its closed subgroups*, Funct. Anal. Appl. **11** (1967), 63–65.

[KZ]   W. Kohnen and D. Zagier, *Fourier coefficients of modular forms of half integral weight*, Math. Ann. **271** (1985), 237–268.

[Kr] L. Kronecker, *Zur Theorie der elliptischen Funktionen*, Werke Vol. 4, 347–495, Vol. 5, 1–132, Leipzig 1929.

[Kl] H. Kloosterman, *On the representation of numbers in the form* $a x^2 + b y^2 + c z^2 + d t^2$, Acta Math. **49** (1926), 407–464.

[Ku] N.V. Kuznetsov, Math. Sbornik III, 153, No. 3 (1980), 334–383.

[L] R. Langlands, *Problems in automorphic forms*, in Springer Lecture Notes **170** (1970), 18–86.

[Li] J. Li, *Poincaré series on* $SU(n,1)$, Preprint 1988.

[LiPS] J. Li, I. Piatetski–Shapiro, and P. Sarnak, *Poincaré series on* $SO(n,1)$, Proc. Indian Acad. Sci. **97** (1987), 231–237.

[Li1] Y.V. Linnik, *Ergodic properties of algebraic number fields*, Springer-Verlag 1968.

[Li2] Y.V. Linnik, *Additive problems and eigenvalues of modular operators*, Proc. I.C.M. Stockholm (1962), 270–284.

[Lo] V. Losert and H. Rindler, *Almost invariant sets*, Bull. London Math. Soc. **13** (1981) No. 2, 145–148.

[LPS1] A. Lubotzky, R. Phillips, and P. Sarnak, *Hecke operators and distributing points on* S^2 *I and II*, Comm. Pure and Appl. Math. **39** (1986), 149–186 and **40** (1987), 401–420.

[LPS2] A. Lubotzky, R. Phillips, and P. Sarnak, *Ramanujan conjecture and explicit construction of expanders*, Proc. STOC 86 (1986), 240–256.

[LPS3] A. Lubotzky, R. Phillips, and P. Sarnak, *Ramanujan graphs*, Combinatorica **8** (1988), 261–277.

[Mal] Malisev, *On the representations of integers by positive definite forms*, Mat. Steklov **65** (1962).

[Ma1] G.A. Margulis, *Some remarks on invariant means*, Monatsch. Math. **90** (1980), 233–235.

[Ma2] G.A. Margulis, *Explicit constructions of concentrators*, Problemy Inf. Trans. **9** (1973), 325–332.

[Ma3] G.A. Margulis, *Arithmetic groups and graphs without short cycles*, sixth Inter. Symp. on Infor. Theory, Tashkent 1984, Abstracts, Vol. I, 123–125.

[Ma4]   G.A. Margulis, *Explicit group theoretic constructions of combina-torial schemes and their applications for construction of expanders and concentrators*, Journal of Problems of Information Transmission (1988).

[Me]   J.F. Mestre, *Courbes de Weil de conducteur 5077*, C. R. Acad. Sci. Paris **300** (1985), 509–512.

[Mey]   G. Meyerson, *Dedekind sums and uniform distribution*, Journal of Number Theory **28** (1988), 233–239.

[Mo]   L. Mordell, *On Mr. Ramanujan's empirical expansions of modular functions*, Proc. Cambridge Phil. Soc. **19** (1917).

[O]   A. Ogg, *Modular forms and Dirichlet series*, W.A. Benjamin, New York 1969.

[Pe]   H. Petersson, *Theorie der automorphen Formen beliebig reeler Dimension und ihre Darstellung durch eine neue Art Poincaré–Reihen*, Math. Ann. **103** (1939), 369–436.

[Pf]   W. Pfetzer, *Die Wirkung der Modulsubstitutionen auf mehrfache Theta–Reihen zu quadratischen Formen ungerader Variablenzhal*, Arch. Math. **6** (1953), 448–454.

[Pi1]   N. Pippenger, *Super concentrators*, SIAM Journal Comp. **6** (1977), 298–304.

[Pi2]   N. Pippenger, Private communication 1988.

[Pin]   M. Pinsker, *On the complexity of a concentrator*, Proc. seventh Int. Teletraffic Conf. (1973), p. 318.

[Pol]   J. Polchinsky, *Evaluation of the one loop string path integral*, UTTG–13–85 (1985).

[R1]   S. Ramanujan, *On certain arithmetical functions*, Trans. Cambridge Phil. Soc. XXII, No. 9, (1916), 159–184.

[R2]   S. Ramanujan, *On certain trigonometrical sums and their applications to the theory of numbers*, Trans. Cambridge Phil. Soc. XXII, No. 13, (1918), 259–276.

[Ra]   R. Rankin, *Modular forms and functions*, Cambridge Univ. Press 1977.

[Ro]   I. Rosenblatt, *Uniqueness of invariant means for measure–preserving transformations*, Trans. A.M.S. **265** (1981), 623–636.

[Ru]   W. Rudin, *Invariant means on L^∞*, Studia Math **44** (1972), 219–227.

[Sar]   P. Sarnak, *Determinants of Laplacians*, Comm. Math. Phys. **110** (1987), 113–120.

[Sa]   I. Satake, *Spherical functions and the Ramanujan conjecture*, Proc. A.M.S. Symp. Vol. 9 (1967), 258–364.

[Sc1]   B. Schoenberg, *Das Verhalten von mehrfachen Thetareihen bei Modulsubstitutionen*, Math. Ann. **116** (1939), 511–523.

[Sc2]   B. Schoenberg, *Elliptic modular functions*, Springer-Verlag, New York 1974.

[Se]   A. Selberg, *On the estimation of Fourier coefficients of modular forms*, Proc. Symp. Pure Math. **8** (1965), 1–15.

[Ser]   J.P. Serre, *A course in arithmetic*, Springer-Verlag 1973.

[Sh]   G. Shimura, *On modular forms of half integral weight*, Ann. Math. **97** (1973), 440–481.

[Si1]   C.L. Siegel, *Lectures on quadratic forms*, Tata Institute, Bombay 1967.

[Si2]   C.L. Siegel, *Über die Classenzahl quadratischer Zahlkörper*, Acta Arith. **1** (1935), 83–86.

[Su]   D. Sullivan, *For $n > 3$ there is only one finitely additive rotationally invariant measure on the n-sphere defined on all Lebesgue measurable subsets*, Bull. Amer. Math. Soc. **4** (1981), 121–123.

[Sz]   G. Szegő, *Orthogonal polynomials*, A.M.S. Coll. Publ., Vol. XXIII, 1939.

[Tan]   R.M. Tanner, *Explicit concentrators from generalized n-gons*, SIAM J. Alg. Disc. Methods **5** (1984), 287–293.

[Tar]   A. Tarski, *Algebraische Fassung des Mass Problems*, Fund. Math. **31** (1938), 47–66.

[Ti]   E.C. Titchmarsh, *The theory of functions*, second edition, Oxford University Press, 1939.

[V]   I. Vardi, *A relation between Dedekind sums and Kloosterman sums*, Duke Math. Journal **55** (1987), 189–197.

[Wal]   J.L. Waldspurger, *Sur les coefficients de Fourier des formes modulaires de poids demi–entier*, J. Math. Pures Appl. **60** (1981), 375–484.

[Wa]   G. Watson, *A treatise on the theory of Bessel functions*, Cambridge Univ. Press 1922.

[We1]   A. Weil, *On some exponential sums*, Proc. Nat. Acad. Sci. **34** (1948), 204–207.

[We2]   A. Weil, *Sur certains groupes d'opérateurs unitaires*, Acta Math. **111** (1964), 143–211.

Index